Slope Engineering

Edited by Ali Ismet Kanlı

Published in London, United Kingdom

IntechOpen

Supporting open minds since 2005

Slope Engineering
http://dx.doi.org/10.5772/intechopen.82508
Edited by Ali Ismet Kanlı

Contributors
Zahid Ur Rehman, Sajjad Hussain, Bushra Nawaz, Noor Mohammad, Akhtar Gul, Masagus Ahmad Azizi, Irfan Marwanza, Afiat Anugrahadi, Muhammad Kemal Ghifari, Yıldırım İsmail İsmail Tosun, Nixon Alexander Correa-Munoz, Carol Andrea Murillo-Feo, Mincheol Park, Dr C Prakasam, Aravinth R, Varinder S Kanwar, B Nagarajan, Gaurav Singh, Raj Kumar, Dinesh Jinger, Dinesh Dhakshanamoorthy, Akshay Kumar Kumar Jha, M.R. Madhav, Ali Ismet Kanlı, Aissa Benselhub, Adelghani Brikat, Adel Aissi, Oussama Kessal, Heuisoo Han, Yoonhwa Jin

First published in London, United Kingdom, 2021 by IntechOpen
IntechOpen is the global imprint of INTECHOPEN LIMITED, registered in England and Wales, registration number: 11086078, 5 Princes Gate Court, London, SW7 2QJ, United Kingdom
Printed in Croatia

British Library Cataloguing-in-Publication Data
A catalogue record for this book is available from the British Library

Additional hard and PDF copies can be obtained from orders@intechopen.com

Slope Engineering
Edited by Ali Ismet Kanlı
p. cm.
Print ISBN 978-1-83962-923-5
Online ISBN 978-1-83962-924-2
eBook (PDF) ISBN 978-1-83962-946-4

For EU product safety concerns:
IN TECH d.o.o., Prolaz Marije Krucifikse Kozulić 3, 51000 Rijeka, Croatia, info@intechopen.com or visit our website at intechopen.com.

Meet the editor

Prof. Dr. Ali Ismet Kanlı received his undergraduate degree in 1989 from Istanbul University, Faculty of Engineering, Department of Geophysical Engineering. In 1994, he graduated from Istanbul University, Institute of Science with an MSc, and in 1998, he completed his doctorate at the same institute. He began his academic career in 1992 as a research assistant in Istanbul University, Faculty of Engineering, Department of Geophysical Engineering, Division of Applied Geophysics. He became an assistant professor in 2001, an associate professor in 2010, and a full professor in 2016 for the same division. Dr. Kanli is the head of the Applied Geophysics Division of the Geophysical Engineering Department. He has carried out and directed many international and national projects and has several national and international scientific publications to his credit. He is an editorial board member of ten international journals and a reviewer for many international and national journals. He has also been a referee in many international and national projects. Dr. Ali is a member of three national and eight international scientific associations. His areas of scientific interest include applied and near-surface geophysics, engineering and environmental geophysics, engineering seismology, exploration seismology, structural geophysics, earthquake engineering, geotechnical geophysics, borehole geophysics and well logging, alternative energy, and geothermal exploration.

Contents

Preface

This book discusses slope engineering, which involves the study and analysis of slope instability problems to safeguard the public and environment as well as the lifelong serviceability of any associated and adjacent structures or infrastructures. Landslides and other gravity-stimulated mass movements are important and costly problems of continual concern for slope engineers. These engineers must pay particular attention to geology, surface drainage, groundwater, and the shear strength of soils in assessing slope stability. Slope engineering projects are frequently located on or near the sloping ground, potentially subject to various kinds of instability such as slides, flows, and falls. These failures produce extensive property damage and occasionally result in loss of life. Therefore, it is frequently necessary to evaluate existing and proposed slopes to assess their stability.

Chapters in this volume cover such topics as remote sensing in landslide detection, tunnel face stability, stability analysis and maintenance of cut slopes, design techniques in rock and soil engineering, statistical models for landslide risk mapping, slope stability analysis in open-pit mines, ecological engineering for slope stabilization, and asphalt-stabilized strengthening in open-pit coal mining.

Ali Ismet Kanlı Ph.d.
Professor of Geophysics,
Faculty of Engineering,
Department of Geophysical Engineering,
Istanbul University-Cerrahpasa,
Turkey

Chapter 1

Design and Construction for Tunnel Face Stability: Theoretical and Modeling Approach

Adel Aissi, Abdelghani Brikat, Ali Ismet Kanlı, Aissa Benselhoub and Oussama Kessal

Abstract

Tunneling is considered to be among the most important projects in all countries worldwide. However, interspersed, some tunnels rise to problems of instability during excavation. This chapter is a case study of the tunnel of "Djebel El Kantour" which is part of the East–West Algerian Highway. Face stability is the most critical problems that affect the subject of our research. This study is carried out via analytical and numerical methods based on the instability relationship, characteristics of the ground and the geometry of the tunnel, to draw conclusions and recommendations for overcoming this problem.

Keywords: replace stability, tunnel face, convergence, confinement, soil characteristics, excavation

1. Introduction

The instability problem of the tunnel face has occurred during the construction of several tunnels in the world (STEFANO Tunnel 1984, TASSO Tunnel 1988, VASTO Tunnel 1991, recognition gallery of St Martin-de-La Porte 2001). This problem has been reported many times in Algeria (Algiers Metro Tunnel 2000, Djebel El Kantour Tunnel in Skikda 2010 and Djebel El Ouhach Tunnel in Constantine 2011). Several studies have been conducted in this regard: [1–3], and various stabilization techniques of tunnels face were used:

1. Fiberglass method of FIT (injection tube) was applied for the first time in 1988 HST Link Roma – Fiorenze(Italy);

2. The technique of shotcrete.

The tunnel face is located at kilometer point (KP 230 + 586.5), which is located inside the tunnel of «Djebel El Kantour". The main problem in this tunnel is the instability of the face which was reported since the beginning of the project, especially in northern the tunnel because of two key factors: the quality of the ground (Marly sandstone clay) and low coverage. In this research finite element simulation were conducted using the software ANSYSE in order take into account the staged construction technique for better estimation the vertical and longitudinal deformations, as well as the failure mechanisms to the front of the tunnel face.

To solve this problem, many techniques have been used to stabilize the face, but the technique did not yield the expected results. Therefore, the prime contractor applied a new technique called the Fiberglass Technique (FIT). However, it proved inadequate in this case, due to its high cost and the limited effectiveness. This is mainly due to the poor mastery of the excavation method, which in turn, was not suitable to this type of rocks. In this study, we try to demonstrate, via a numerical modeling tool, the relationship between the attack section and the deformation field on the Tunnel face.

2. Case study. Djebel El Kantour tunnel

T4 tunnel is located in the north-east of the department of Constantine. It crosses Djebel El Kantour from south to north with a total length of 2500 m. Its cover is higher than 15 m, and reaches 224 m in maximum points (**Figure 1**).

The section of the tunnel was chosen according to the geometric characteristics, the geological and geotechnical data tests performed on the ground in question as well as the height of the cover. To take into account the natural conditions of the surrounding terrain, an arched profile has been adopted to ensure the stability of the structure and the service conditions during the constructions process.

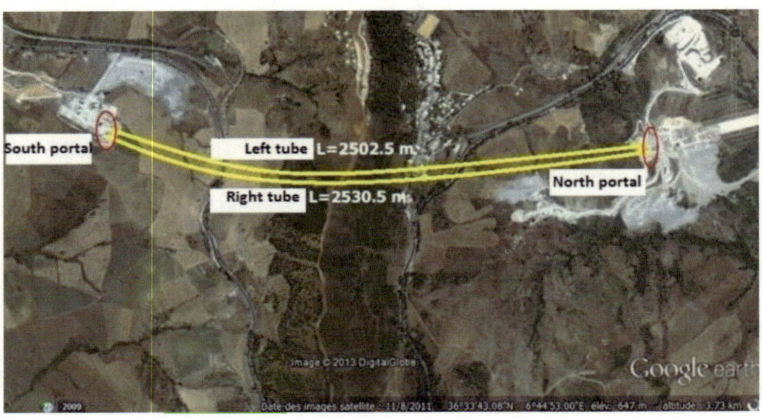

Figure 1.
Position of the tunnel T4.

3. Geological and geotechnical parameters of the study

The design of the tunnel was performed on the basis of a geological and geomechanical survey conducted from recent geotechnical survey includes the following investigations:

- Geological surveys conducted by geologists' experts;
- A recognition campaign by core drilling, in situ and laboratory tests.

It noted that any study must lead to the acquisition of the following information with the maximum possible degree of reliability, considering the wide range of technical resources currently available in the geological survey field:

- Structural geological and hydrogeological conditions and the natural stress state of the ground to be tunneled;
- The physical characteristics, the strength and deformability of the geological bodies affected by the excavation:
- The hydrogeological conditions in the rock mass.

Geology of the area crossed by the tunnel is essentially of Cretaceous age (Telliennes Tablecloths) and consists of marl and limestone in the form of strongly folded and sheared blocks (**Figure 2**). These are covered by Quaternary deposits consisting of clays, silts and conglomerates.

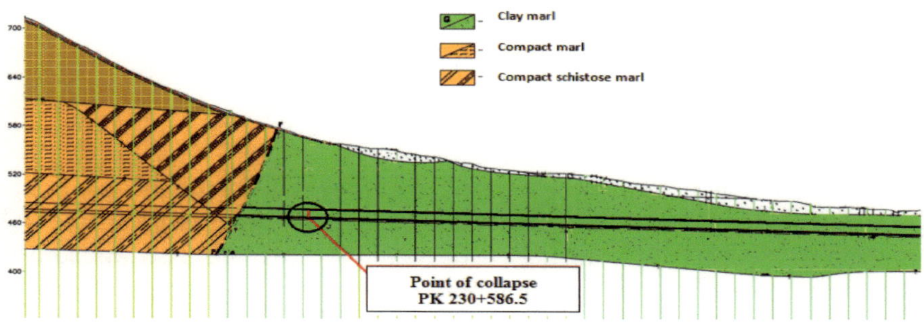

Figure 2.
Plan and geological section of the T4 tunnel and the collapse point.

N	Parameters	Coefficients	Rating
1	Resistance to uniaxial compression	R0 $(R0 < 0.1 MPa)$	0
2	Rock Quality Designation (RQD)	(115–3.3Jv) 0–25 very poor	3
3	Spacing of discontinuities	$< 600 mm$	5
4	Conditions of discontinuities	Slickensided surfaces; 3–10; 10–20; 0.1–5 $> 5mm$;< 5 mm; highly altered	6.5
5	Groundwater	Damp	10
6	Rating adjustment for discontinuity orientations	Digging against the dip	−12
Total rating			12.5

Table 1.
Results according to the RMR classification system in the PK 230 + 586.5.

Total rating	<21
Class	V
Average stand up time	30 min for 1 m span
C « KPa »	<100
φ(°)	<15
Description	Very poor rock

Table 2.
Total-rating results.

γ (kN/m³)	20
E (MPa)	200
C (kPa)	50–160
φ(°)	25

Table 3.
Geotechnical parameters.

The design of the tunnel was carried out on the basis of geological and geotechnical studies. The results of RMR classification are presented in **Table 1**.

The table below (**Table 2**) shows the value of the rock classification (Rock Mass ratings) determined after application of the Total rating.

In our case, the rocks are of marl-clay-sandstone type (**Table 3**).

4. Collapse problem of the tunnel face at PK 230 + 586.5

This tunnel consists of two tubes spaced 22 m. The problem of collapse occurred on the southern side of the tunnel in the right tube. The RMR classification results obtained on the Southern side during the day that preceded the crisis was similar to the classification results of the opposite side of the tunnel, which also suffers from the same problem of instability, but with a technique linked to the soil stabilization.

As we dig in the initially stable soil, the preexisting stress state has changed. Indeed, the stress on the excavation contour vanishes: the decompression phenomenon. This change in the stress state appears only in an area surrounding the Tunnel face: the influence area of the face. It extends over a length towards the front edge which is of the same order of magnitude as the diameter of the tunnel according to the measurements performed on several displacement starts [4, 5].

The usual methods for calculating the tunnel's, Tunnel face stability are resulting from experimental studies [6], extrusion testing in laboratory [7] Semi-empirical and theoretical which mainly the approach of calculating the rupture [8, 9].

In our case, the experiment shows that the ruptures of the Tunnel face can mobilize important volumes of ground.

The first systematic studies on the face instability of the tunnels dag in the soft soil carried out by [10] were used to characterize the stability conditions starting from a stability parameter coefficient (**Figure 3**).

The stability coefficient N is defined in [10].

$$N = \frac{\sigma\_s - \sigma\_T}{Q\_u} + \frac{\gamma.(C+R)}{Q\_u} \tag{1}$$

Where;

γ: density of the rock.

Q u: Shearing resistance (**Figure 4**).

The **Figure 5** gives an indication of the relation between the amount of real stability and awaited deformations.

In our case, the parameters required for the calculation of the stability coefficient N are the following:

- The surface load; $\sigma_s = 14.10^5\,\text{Pa}$ and $\sigma_T = 0$;

- The cover C = 25 m;
- The tunnel radius R = 5 m. $\gamma = 2.10^4\ N/m^3$; Qu =3,1.105 Pa;

The obtained result for the stability coefficient is N = 6.45, which means that the tunnel face is unstable.

In our case, the Tunnel face is unstable; therefore, we try to find the value of the pressure of supporting σ_T suitable to be applied in order to decrease this state towards an Elastoplastic deformation.

The interval of σ_T is calculated starting according to the following formula:
N ϵ [2 untill 4]

$$2 \le \frac{\sigma_s - \sigma_T}{Q_u} + \frac{\gamma.(C+R)}{Q_u} \le 4 \qquad (2)$$

$$2 \le \frac{14.10^5 - \sigma_T}{3{,}1.10^5} + \frac{2.10^4.(25+5)}{3{,}1.10^5} \le 4$$

The required σ_T an Elastoplastic deformation the following:

$$13{,}17.10^5\,Pa \ge \sigma_T \ge 6{,}97.10^5\,Pa$$

In this case, the pressure to be exerted on the Tunnel face will lie between the two following values:

$$\sigma_T \epsilon\ [\approx 0.7 \text{ untill} \approx 1]\ (MPa).$$

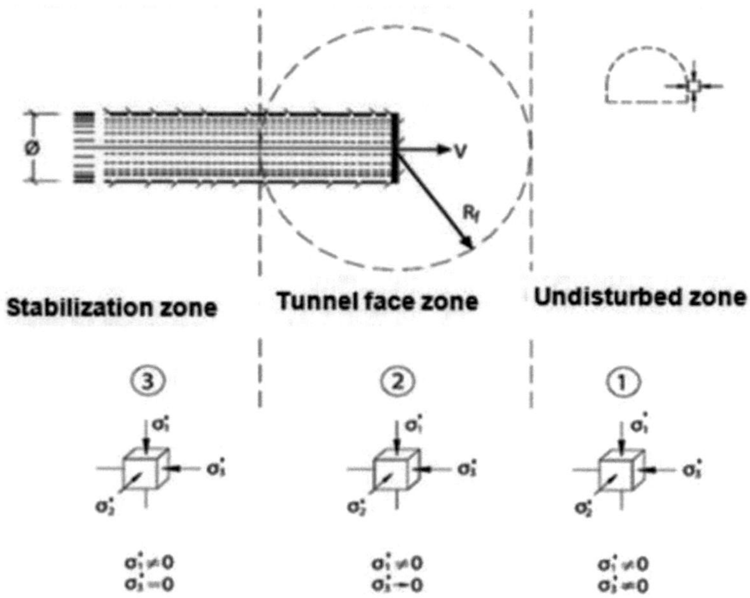

Figure 3.
Formation of three characteristic zones during the digging of a tunnel (Lunardi 1993) [11].

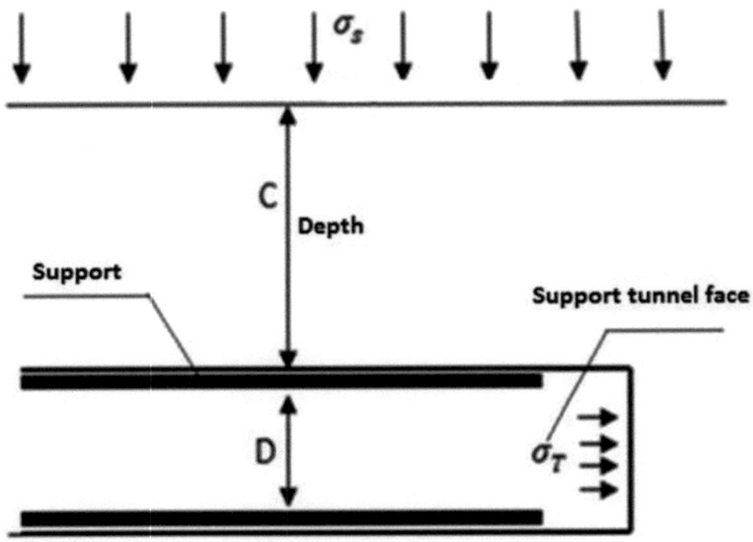

Figure 4.
Tunnel face schematization.

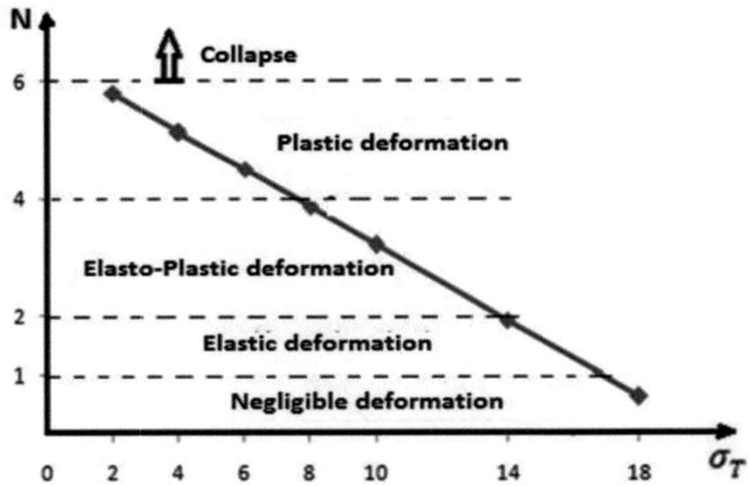

Figure 5.
Relation between the stability coefficient and the face supporting according to [12].

The state of stress in the ground is considerably greater than the strength properties of the material even in the zone around the face. For this consideration based for the results of the diagnosis phase, the techniques to be applied for the application of the supporting pressure on the Tunnel faces are as follows:

- Shotcrete;
- Fiber glass bolting.

Both methods assure the rigidity of the core of ground ahead of the face, and therefore the conditions of stability in that ground, have a decisive effect on that deformation response and determine how an arch effect is triggered and consequently the tone of the stress–strain response in the whole tunnel.

On the other hand, they are difficult to apply this theoretical approach in the domains of soft rocks, flysch and soils, give insufficient consideration to the effects of natural stress states and the dimensions and geometry of an excavation on the deformation behavior of a tunnel and fail to take account of new constructions systems [13].

However, it does not give adequate consideration to the construction stages and therefore it does not constitute a fully integrated method of design and construction. To do this, our analysis of the deformation response continued using the numerical modeling which is able to consider stress states in the ground that are not of the hydrostatic type, which take due account of gravitational loads and which also calculate the effects which the various construction stages have on the statics of a tunnel by simulating the real geometry of lining structures and the sequence and the distance.

5. Numerical analysis

In engineering practice, different design methods tend to be used; in this study, advanced numerical modeling was used due to its ability to predict vertical and longitudinal deformations; as well as the failure mechanisms at the front of the tunnel face [14, 15]. It can indeed be used to simultaneously take into account constraints and anisotropic materials, tunnel advance stages and any pre-containment and cavity containment intervention. In this work, the use of a calculation code through finite element method according to the execution situation [16–18]. The numerical parameter used in the simulation as already mentioned in the **Table 3** resulting from the geotechnical investigation of the zone in question, where the behavior criteria used in the simulation is Drucker-Prager criterion, which as a generalization of the Mohr–Coulomb criterion for soils. The criterion is based on the assumption that the octahedral shear stress at failure, it depends linearly on the octahedral normal stress through material constants. The results indicate that the action of the surrounding terrain on the tunnel based on the attack section R. The main input parameters are the mesh network of elements which determines the domain to which the analysis applies, the geomechanical properties of each element, the surrounding conditions and the loads acting (**Figure 6**).

Numerical simulations allowed obtaining practical results of the radial and longitudinal displacements in the figure and table below:

5.1 Longitudinal displacement

The following diagram shows the extrusion of different attack section based on the distance in front of the face (**Figure** 7).

5.2 Radial displacements vertical displacements

The vertical and longitudinal deformations and changes of the critical zone that occur in the tunnel are linked to attack section R. When the maximum value of the attack section "R" is equal to 5 m, the corresponding Maximum Vertical Displacement (Ux) is equal to 0.53 m, and Longitudinal Deformation in front of the Tunnel face can reach a maximum of 65 m.

In the case where the radius R = 3.5 m attack, the maximum vertical displacement (Ux) is 0.2 m. So, it is 3 times less than the previous case, and similarly for Longitudinal Deformation in front of the Tunnel face which does not exceed 50 m. it is shown that simulation results are consistent with the observed extent and those obtained in literature. (**Figure 8**).

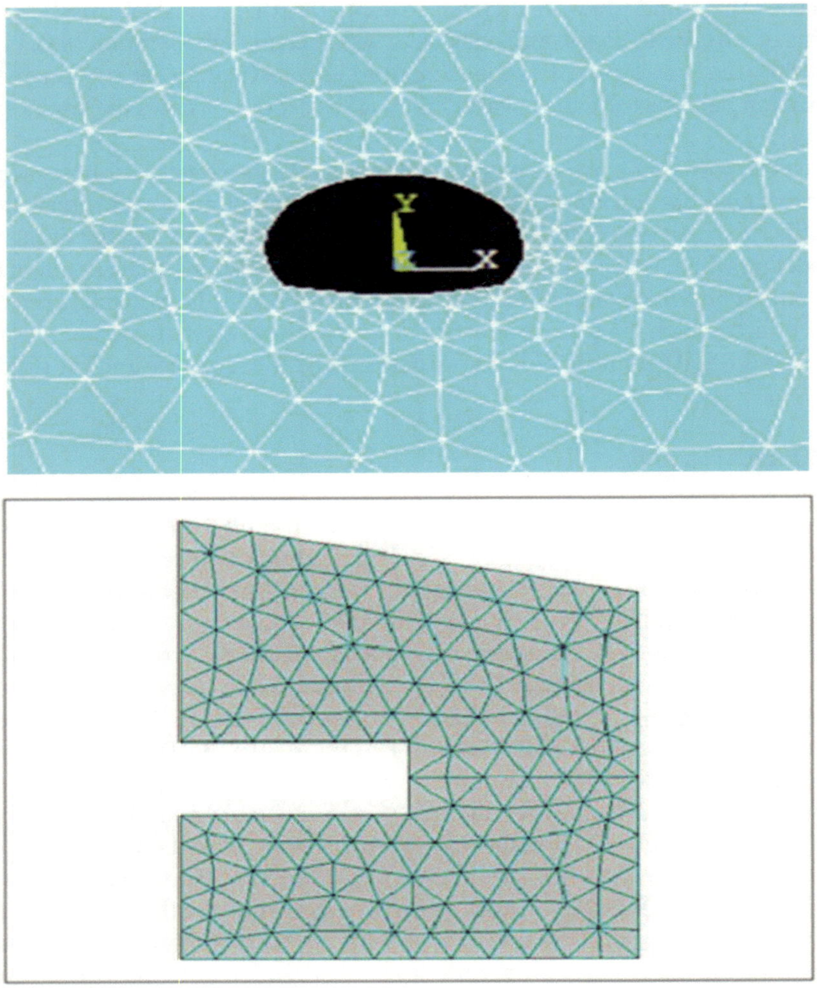

Figure 6.
Model and meshing.

Figure 7.
Uz displacement curves based on the attack section R.

5.3 Horizontal displacements

We studied the horizontal displacement of the solid rock in vertical section for all cases of attack sections.

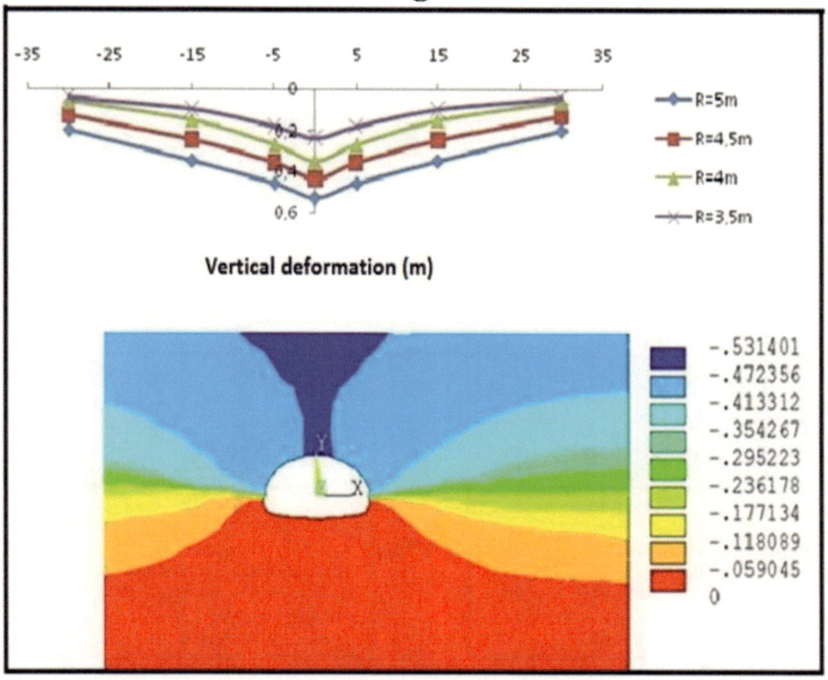

Figure 8.
Ux vertical displacement curves based on the attack section R.

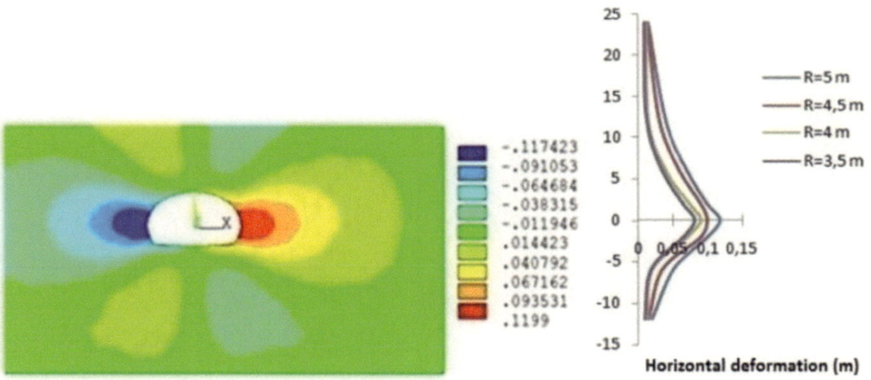

Figure 9.
Uh horizontal displacement curves based on the attack section R.

The (**Figure 9**) shows the horizontal displacements along a vertical section for various attack sections R.

The maximum horizontal displacement Uh is the edge of the excavation to tend towards zero displacement at a considerable distance from the tunnel. With a difference between the displacements, values of every driving section R compared to another.

6. Conclusion and recommendations

To ensure the stability of the Tunnel face and minimize its stress concentrations and deformations; the exploitation of this study's results with the inclusion of

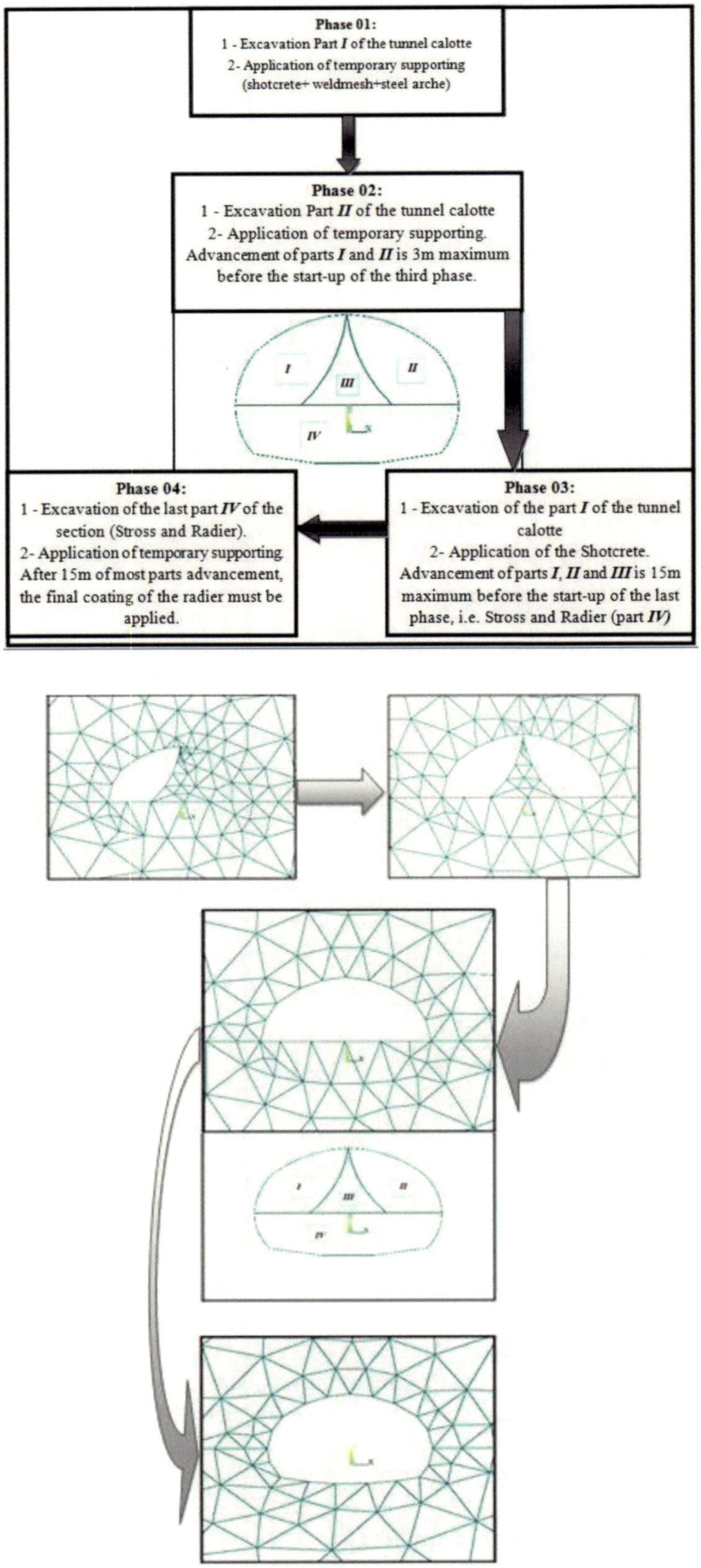

Figure 10.
Staged construction with divided section.

geometric and geological conditions traversed by the tunnel, we recommend the following points:

- The change in the advancement method in divided sections when terrain features are insufficient to ensure the necessary stability to the face. This method is adopted when the excavation cannot be performed in full section.

- The chart below (**Figure 10**) shows the progress detailed in the proposed divided section.

Applying pressure on the supporting σ_T on the Tunnel face to reduce its deformation, of which values interval should be:

$$\sigma_T \epsilon \left[\approx 0{,}7 \text{ untill} \approx 1{,}3\right] \text{ (MPa)}.$$

It is suggested that the technique proposed maybe used for construction considerations under complication geological condition which satisfactory effect in engineering practice.

Author details

Adel Aissi[1], Abdelghani Brikat[2], Ali Ismet Kanlı[3], Aissa Benselhoub[4*] and Oussama Kessal[5]

1 Mohamed Boudiaf University, M'Sila, Algeria

2 Badji Mokhtar University, Annaba, Algeria

3 Istanbul University-Cerrahpasa, Istanbul, Turkey

4 Environmental Research Center, Annaba, Algeria

5 Mohammed El-Bachir Ibrahimi University of Bordj Bou Arréridj, El-Anasser, Bordj Bou Arréridj, Algeria

*Address all correspondence to: benselhoub@yahoo.fr

References

[1] LombardiG : La révision dans la construction des tunnels. Géologie et mécanique des roches,: COLLOQUE GEOLOGIE DE L'INGENIEUR; 1974 LIEGE, p 149-162.

[2] Laca E, PanetM: Application du calcul à la rupture à la stabilité du front de taille d'un tunnel, Revue française de géotechnique, n° 43, 5-20.1988.

[3] Pietro L : convergence-confinement ou extrusion-préconfinement, Conférence Tenuta (Parigi),129-145.1998.

[4] Pietro L: Design and Construction of Tunnels, Analysis of Controlled Deformation in Rocks and Soils. Springer-Verlag Berlin Heidelberg, 3-45. 2008.

[5] TrompilleV: Etude expérimentale et théorique du comportement d'un tunnel renforcé par boulonnage frontal, Thèse de doctorat, INSA de Lyon, 19-20.2003.

[6] Chern JC, Shiao FY, Yu C.W:An empiricalsafetycriterion for tunnel construction: Proceedings of the Regional Symposium on Sedimentary Rock Engineering. Taipei, 222-227. 1998.

[7] Centre d'études des tunnels, Juillet. Dossier pilote des tunnels. France, 21-27. 1998.

[8] Broere W: Tunnel Face Stability and New CPT Applications. Ph.D Thesis – Technical University of Delft, 5-30. 2001.

[9] Davis E H, Gunn M J, Mair RJ, and Seneviratne H N: The stability of shallow tunnels and underground openings in cohesive material. Géotechnique, 30(4):397-416, 1980.

[10] Broms B, Bennermark H: Stability of clay at vertical openings', Journal of the SoilMechanics and Foundations Division, ASCE, SM1, 71-94. 1967.

[11] Pietro L: Fibre-Glass tubes to stabilize the face of tunnels in difficult cohesive soils Seminar on "The application of fibre reinforced plastics (FRP) in civil structural engineering" Bolgne, 107-165.1993.

[12] Zienkiewicz O C, Taylor R L : La méthode des éléments finis : formulation de base et problèmes linéaires, AFNOR technique, Paris. 610-625.1991.

[13] Pietro L: Design and construction of tunnels. editors. Springer-Verlag Berlin Heidelberg. p. 23-587. 2008. DOI: 10.1007/978-3-540-73875-6.ch2

[14] T.Baumann R,SternathJ, Schwarz : Face support for tunnels in losse grouds in ; world tunnel congress, tunnel for people 317-324;1997; Vienna.

[15] Huang M S,Jia C Q: Stability analysis of soil slopes subjected to unsaturated transient seepage, chinese journal of geotecnical engineering 28(2), 202-206, 2006.

[16] DHATT G., TOUZOT G Présentation de la méthode des éléments finis, 2éme Edition MALOINE Paris.1984.

[17] DHATT G., TOUZOT G : Présentation de la méthode des éléments finis, 2éme Edition MALOINE Paris.1984.

[18] SABONNIERE J C, COULOMB J.C : Eléments finis et CAO, Edition HERMES Paris, 210p.1986.

Chapter 2

Geosynthetic Reinforced Embankment Slopes

Akshay Kumar Jha and Madhav Madhira

Abstract

Slope failures lead to loss of life and damage to property. Slope instability of natural slope depends on natural and manmade factors such as excessive rainfall, earthquakes, deforestation, unplanned construction activity, etc. Manmade slopes are formed for embankments and cuttings. Steepening of slopes for construction of rail/road embankments or for widening of existing roads is a necessity for development. Use of geosynthetics for steep slope construction considering design and environmental aspects could be a viable alternative to these issues. Methods developed for unreinforced slopes have been extended to analyze geosynthetic reinforced slopes accounting for the presence of reinforcement. Designing geosynthetic reinforced slope with minimum length of geosynthetics leads to economy. This chapter presents review of literature and design methodologies available for reinforced slopes with granular and marginal backfills. Optimization of reinforcement length from face end of the slope and slope - reinforcement interactions are also presented.

Keywords: slopes, geosynthetics, reinforcement, optimization of length, marginal soils, steepening

1. Introduction

Landslides in slopes and failures of embankment and cut slopes lead to loss of life and property. Several factors, natural and manmade, such as heavy rainfall, unplanned construction, deforestation, restricting waterways of rivers and their tributaries are major causes for instability of slopes. Factors controlling stability of natural slopes are type of soil, environmental conditions, groundwater, stress history, rainfall, cloud burst, earthquakes, etc. Landslide mortality rate exceeds one per 100 km^2 per year in developing countries like India, China, Nepal, Peru, Venezuela, Philippines and Tajikistan [1, 2]. Factors that cause man-made slope failure are very different and could be due to inadequate design, improper backfill, poor construction, etc. The repair of failed slope involves removal of debris and reinstatement of slope with free draining material. Restoration of the slide with geosynthetics can be simpler, faster and economical. Designing slopes with Geosynthetics has several advantages (Simac [3]), e.g., reduced land requirement, additional usable area at toe of slope, use of available on-site soil, reduced transportation costs of select fill or export costs of unsuitable fill, steeper slopes, elimination of concrete facing, and facilitation of natural vegetation for sustainable development.

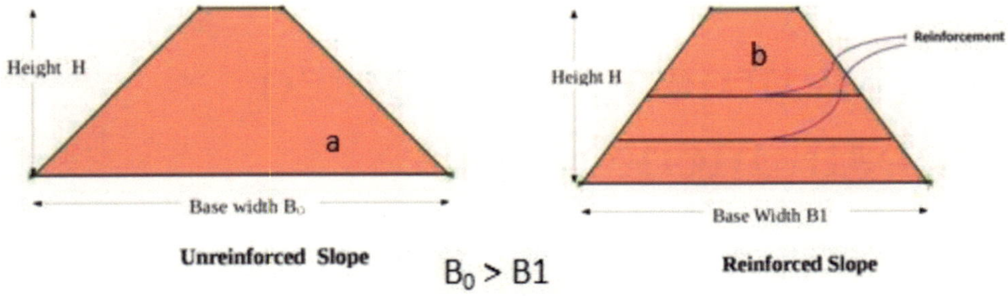

Figure 1.
Effect of geosynthetic reinforcement on geometry of embankment: (a) flat and (b) steep slopes.

Embankments are built with engineered fills. Geosynthetics facilitate reduction of earthwork volume by altering the geometry of the embankment (**Figure 1**) and even allow use of marginal soil.

In reinforced soil, conventionally free draining material is specified for backfill due to its high strength and draining properties. The cost of fill material is about 40% of the total cost of the structure [4]. If marginal soil is used instead it could be more economical. Apart from economics, technical factors like esthetics, reliability, simple construction techniques, good seismic performance and ability to tolerate large deformations without structural distress have enhanced the acceptance and use of geosynthetics as reinforcing material.

2. Case histories

2.1 Tallest geosynthetic reinforced slope

One of the tallest geogrid reinforced green faced slope 1H:1 V and 74 m high (**Figure 2**) is airport runway extension in Charleston, West Virginia [5]. **Figure 3** depicts a schematic of the reinforced slope.

The high angle of shearing resistance of 36° of onsite soils provided two significant benefits viz., no need to import borrow soil and minimum required embedment length. However, there was a partial failure of this slope which was restored subsequently.

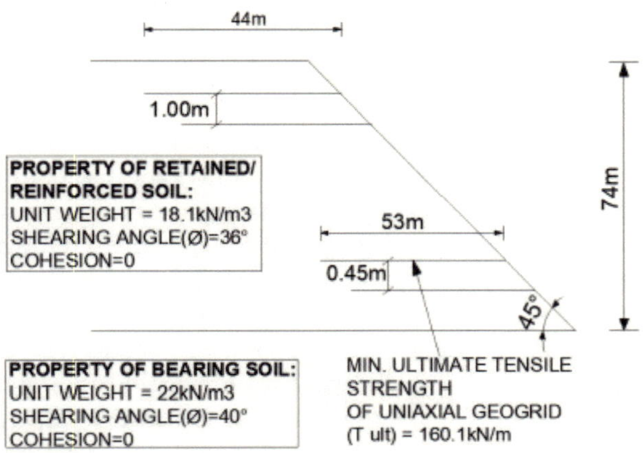

Figure 2.
Schematic of reinforced embankment for Yeager runway extension (redrawn based on [5]).

Figure 3.
Reinforced slope Yeager runway extension (after [5]).

2.2 Geogrid reinforced embankment with steep side slopes

For a highway project in Brampton, Ontario, property acquisition costs and other problems necessitated the design of 7 m high embankment with 1H:1 V steep side slopes. Cost–benefit studies showed that steep side slopes reinforced with synthetic tensile elements were considerably cheaper than the other alternatives. Design and construction of 7 m high embankment with slopes using geogrids (**Figure 4**) as an alternative to rock fill embankment with side slopes of 1.25H:1 V requiring additional land has been covered by Devata [6]. Reinforced earth slope was found to be most economical (**Table 1**). The total cost of 1H:1 V reinforced slope is the least among all the alternatives making it the most economical option.

2.3 Composite soil reinforcement system for very high and steep fills for Sikkim airport

For runway construction of airport at Pakyong in north eastern Indian state of Sikkim huge cutting of earth and its filling on the valley side was required to form a level platform to provide runway of 1820 m x 150 m and other related

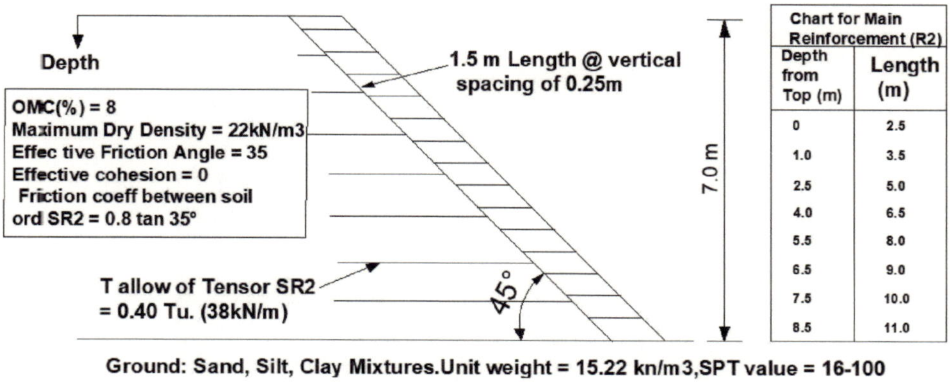

Figure 4.
Schematic of reinforced slope for highway extension (sketch based on [6]).

Construction Method	Construction cost $M	Property Cost $M	Total Cost $ M
Reinf. Concrete Wall	1.52	Nil	1.52
Reinforced Earth Wall	1.42	Nil	1.42
1.25H:1 V Rockfill	0.93	0.30	1.23
1H:1 V Reinf. Slope	0.48	Nil	0.48
2H:1 V Earth Fill	0.30	0.90	1.20

Table 1.
Cost comparison of different options.

infrastructures over 200-acre area. Retention of fill on the valley side needed construction of retaining structures with heights varying from 30 to 74 m over a length of 1480 m (**Figure 5**).

On hill side, the cut slopes have a height extending up to 100 m. To retain and stabilize this fill of varying height, a composite soil reinforcement system was employed [7]. To make optimum utilization of space available and minimize cost, combination of vertical wall and steep slope has been adopted for construction of retaining structure. Facing elements for the reinforced soil wall comprise of Gabions (**Figure 6**).

Figure 5.
***Sikkim** airport - aerial view (after [7]).*

Figure 6.
Slope face completely covered with vegetation few months after installation (after [7]).

3. Geosynthetics

Geosynthetics are mostly planar products manufactured from polymeric material and used with geomaterials such as soil and rock, as integral part of manmade project or system for better performance, economy, better quality control, rapid installation, cost competitiveness, lower carbon footprint, requirement of smaller parcel of land for embankments, etc.

Geotextiles and Geogrids are used normally for reinforcing embankments or natural slopes either to obtain higher factor of safety or for construction with steep slopes. Allowable Geotextile/Geogrid strength is arrived at using several factors to account for degradation, creep, installation damage, etc. The allowable tensile strength, T_{all}, is

$$T_{all} = T_{ult} / R_{FID} * R_{FCR} * RF_{CBD} \tag{1}$$

Where T_{all} and T_{ult} - allowable and ultimate tensile strengths respectively, R_{FID}, R_{FCR} and R_{FCBD} - reduction factors (all >1.0) for installation, creep and chemical and biological damage respectively. The combined or overall reduction factor is about 2.0 for design.

4. Literature review

Jewell et al. [8], Bonaparte et al. [9] and Verduin and Holtz [10] present design methods for earth slopes reinforced with geotextiles and/or geogrids using limit equilibrium method considering circular or/and bilinear wedges. Leshchinsky and Reinschmidt [11] and Leshchinsky and Boedeker [12] present an approach based on limit equilibrium and variational extremization of factor of safety of multilayer reinforced slope. Schneider and Holtz [13] present a design procedure for slopes reinforced with geotextiles and geogrids for a bilinear surface of sliding, considering porewater pressures and the initial stress conditions in the slope. Jewell [14] presented revised design charts for steep slopes valid for all reinforcement materials. Leshchinsky [15] and Leshchinsky et al. [16] used log-spiral failure mechanism to determine the required reinforcement long term strength. Zhao [17] and Michalowski [18] present kinematic limit analyses solutions for the stability of reinforced soil slopes. Shiwakoti et al. [19] conducted parametric studies to investigate the effect of geosynthetic strength, soil–geosynthetic interaction coefficients, vertical spacing of geosynthetics for soil slope/wall on competent foundation. Baker and Klein [20, 21] modified the top-down approach of Leshchinsky [15] to obtain the reinforcement force needed for a prescribed factor of safety everywhere within the reinforced mass. Han and Leshchinsky [22] present a general analytical framework for design of flexible reinforced earth structures, i.e., walls and slopes. Leshchinsky et al. [23] present a limit equilibrium methodology to determine the unfactored global geosynthetic strength required to ensure sufficient internal stability in reinforced earth structures. Leshchinsky et al. [24] introduced a limit state design framework for geosynthetic reinforced slopes and walls. Leshchinsky and Ambauen [25] present use of upper bound limit analysis (LA) in conjunction with discretization procedure known as discontinuity layout optimization (DLO). DLO-LA is an effective tool for establishing a critical failure mechanism and ensuing stability of the slope without the constraint or assumptions required in LE analysis. Shukla et al. [26] presented a review of design of reinforced slope and covers basic of methods in detail. Gao et al. [27] in their study considered three-dimensional effect on reinforced earth structure stability and to determine the required strength and length of reinforcement using limit analysis. Song et al. [28] proposed new approach based on LE principle to evaluate stability of reinforced slope.

Free draining granular material is used conventionally for reinforced earth slope construction. However cohesive materials have also been used for construction of reinforced slopes in few cases. Very few design guidelines/methods are available for design of reinforced earth slope with marginal soil. Christopher et al. [29] provide design guidance (total stress analysis ignoring the drainage contribution of geocomposite for short term and effective stress analysis considering drainage in the long term) for reinforced soil structures using poorly draining backfills. Naughton et al. [30] improved the design method of Christopher et al. [29] and presented single stage effective stress analysis since excess pore pressure gets dissipated fully before construction of subsequent layers. Clancy and Naughton [31] used design approach of Naughton et al. [30] to design four steep slopes using fine-grained soils as backfill material and provided a method to determine the maximum height of each lift to allow dissipation of excess pore pressures in a 24-hour period for a 10 m high 70° slope. Giroud et al. [32] updated design method of Naughton et al. [30] for reinforced slopes and walls using draining geogrid, with focus on improved determination of the required transmissivity of the same. Naughton et al. [33] conducted a parametric study of design parameters of low permeability fill and concluded that for typical compressibility and consolidation parameters vertical spacing of the reinforcement of 0.5 m is adequate.

Abd and Utili [33] employed limit analysis approach and semi-analytical method for uniform slopes that provide the amount of reinforcement needed as a function of cohesion, c, and angle of shearing resistance, ϕ, of backfill, tensile strength of geosynthetic and of the slope inclination.

5. Design methods

Geosynthetic reinforced slopes are designed to provide internal, external, global and surfacial stability. Surfacial stability determines the requirement of secondary reinforcement to ensure no shallow sloughing. The design process must address all possible failure modes that a reinforced (or unreinforced) slope would potentially experience. The design addresses internal stability (pull out and bond failures) for the condition where the failure surface intersects the reinforcement, external stability (sliding, overturning, bearing failures) for the condition where the failure surface is located outside and below the reinforced soil mass and compound stability for the condition where the failure plane passes behind and through the reinforced soil mass. In order to analyze a reinforced slope the requirements include the slope geometry, external and seismic loading, porewater pressure and/or seepage conditions, soil parameters and properties, the reinforcement parameters and properties, the interaction characteristics of the soil and the geosynthetic. The design of a reinforced soil slope determines the final geometry, the required number, spacing and lengths of reinforcement layers and measures to prevent sloughing or erosion of the slope face.

Methods originally developed for unreinforced slopes have been extended to reinforced slopes accounting for the presence of reinforcement. Methods available for analyzing geosynthetic reinforced soil slopes are (i) Limit equilibrium, (ii) Limit analysis, (iii) Slip line and (iv) Finite element methods.

5.1 Limit equilibrium method

Conventional geotechnical engineering approach to slope stability problems is to use limit equilibrium concepts on an assumed circular or non-circular failure surface and to arrive at a factor of safety. Factor of safety is estimated using moment and/or force equilibrium equations considering the reinforcing effect of geosynthetics. Several limit equilibrium methods have been used in various studies [34–39].

5.2 Limit analysis

Limit analysis is another method for solution of slope stability problems [17, 19, 25, 40–44]. It is based on plasticity theory and can be applied to slopes of arbitrary geometry and complex loading conditions. Using limit theorems, collapse load can be bracketed between lower and upper bounds even if it cannot be determined exactly. Recent approaches that combine finite elements and failure criterion have narrowed the gap between the two bounds.

5.3 Slip line method

Slip line method is based on stress characteristics and based on homogenization of the composite mass and suitable for continuous filament or fiber reinforced soil slopes. Failure criterion for geosynthetic reinforced soil composite was presented by Michalowski and Zhao [40]. Limit loads on geosynthetic reinforced soil slopes using slip line method were given by Zhao [17].

5.4 Finite element method

Finite element method of analysis is generally based on quasi-elastic continuum mechanics approach in which stresses and strains are estimated throughout the mass. In this method both deformation and strength parameters, viz., modulus of deformation (E), Poisson's ratio (υ), cohesion, c, angle of shearing resistance, ϕ, angle of dilation (ψ) are required for design. Alternately, shear strength reduction technique is used for design of slope considering the effect of reinforcement. In this approach no assumption needs to be made regarding nature of failure surface or its location as failure occurs "naturally" through the zones within the soil mass wherein the shear stresses attain values close to the strength of the soil. Details of this approach can be found in the works of Rowe and Soderman [45], Almeida et al. [46], Chalaturnyk et al. [47], Ali and Tee [48] and Griffith and Lane [49]. Software such as Plaxis, FLAC, etc., are available for the analysis by 'Soil Strength Reduction Technique' using FEM.

6. Reinforced embankment slope design - Jewell method

6.1 Jewell et al. design method

Jewell et al. [8] proposed a method of design based on Limit Equilibrium analysis and local check on individual reinforcement spacing for geogrid reinforced embankment slope for slope angles, β, ranging from 30° to 80°. Embankment soil is granular and the crest is horizontal. Length of reinforcement is based on (i) no overstressing of lower layers, (ii) no outward sliding along the interface between soil and reinforcement layer and (iii) no tension on the base. Two-part wedge analysis is used and critical wedge surface for failure is located by varying wedge angles θ_1 and θ_2 (**Figure 7**) and the location of intersection of the two wedge lines, i.e., the wedge point.

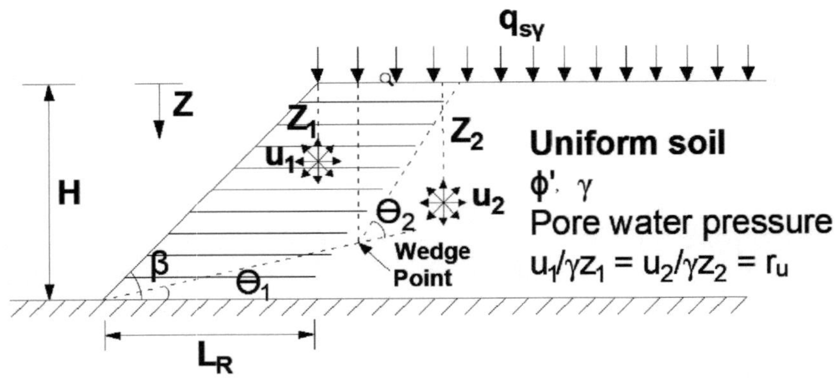

Figure 7.
Steep slope embankment (after [8]).

The interslice forces are assumed to be zero. Design charts calculate total reinforcement force and length of reinforcement in terms of slope angle, β, effective critical state friction angle of soil, ϕ c, and porewater pressure parameter, $r_u = u / \gamma z$. The strength of reinforcement is the strength of geogrid at the end of design life for most severe condition expected during service life. Factor of safety of 1.3 to 1.5 for the slope is achieved by applying the same to the reinforcement strength as well. All reinforcement layers are of equal length. Surcharge and earthquake loads are not included in the analysis.

6.2 Revised design

Jewell [14] revised the previous design for geotextiles and geogrids as reinforcement. Interaction between soil and horizontal reinforcement has been considered in terms of bond coefficient (f_b) which governs the load transfer between reinforcement and soil. The basic philosophy of design is that available stress from the reinforcement exceeds the required stress for equilibrium in soil. Improvements over previous design method are given in **Table 2**.

6.2.1 Design parameters for soil and reinforcement

Allowable Tensile Strength (T_{all}) for reinforcement is chosen such that strain in reinforcement does not exceed 3–5% during design life to ensure satisfactory serviceability. ϕ d = ϕ_c . Porewater pressure to be considered should include the worst condition expected in design life. Design values of ϕ $_d$, r_u, β and H are used to

SN	Parameter	Jewell et al. [8]	Jewell [14]
1	Slope angle (β)	$80° \le \beta \le 30°$	$90° \le \beta \le 30°$
2	Bond coefficient (f_b)	Reinforcement bond angle of friction = 50% design friction angle of soil	$1 \ge fb \ge 0$
3	Direct Sliding coefficient (f_{ds})	Friction resistance to direct sliding = 80% design friction angle of soil	f_{ds} = 0.80 and correction factor applied whenever f_{ds} takes a value less than 0.8.
4	Reinforcing material	Geogrid	Geogrid and Geotextile

Table 2.
Salient improvements over Jewell et al. [14].

determine required earth pressure coefficient, K_{req}, and reinforcement length, L_R/H, from the design charts. Design charts for r_u = 0, 0.25 and 0.50 are shown in **Figures 8** and **9**. Large strain or critical state shearing resistance (ϕ_c) of soil is to be used for design.

6.3 Limit equilibrium (LE) method of stability analysis

Various LE Methods such as Bishop's Simplified, Janbu's Simplified, Spencer, Morgenstern-Price, Janbu Generalized, Sarma, etc., have been developed for slope stability analysis. The problem is considered in two dimension i.e. plane strain case. The primary difference among all these methods lies in the equations of statics considered, which interslice normal and shear forces are included, and the assumed relationship between the interslice forces. **Tables 3** and **4** summarize the conditions for some of the common methods of stability analysis.

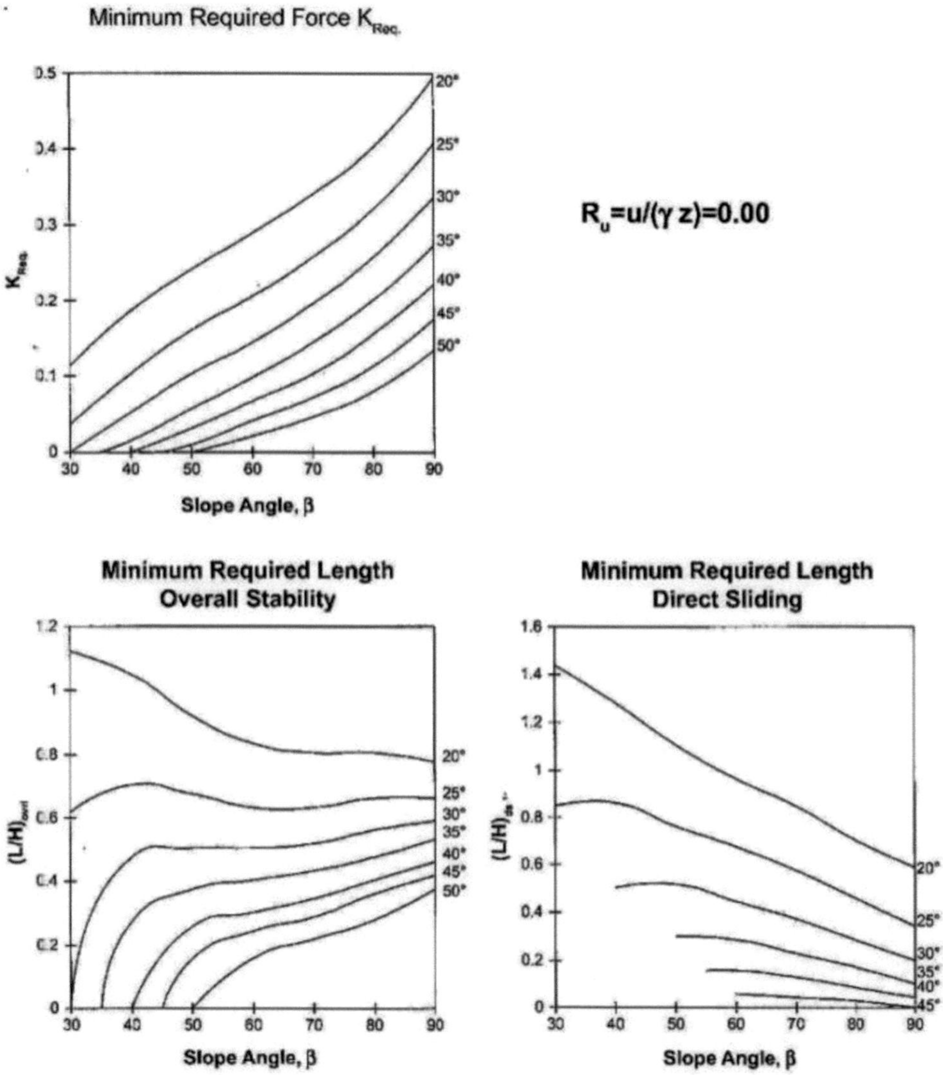

Figure 8.
Design chart for r_u = 0 after Jewell [14].

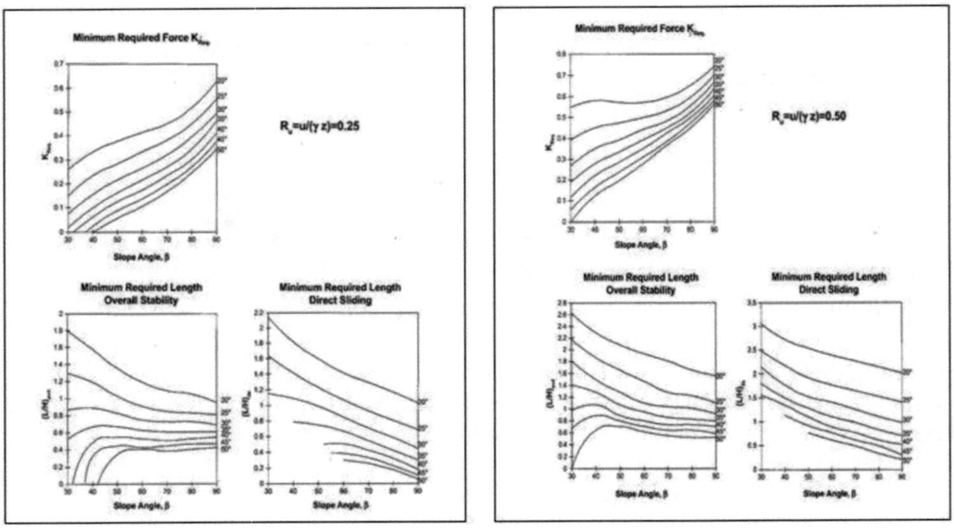

Figure 9.
Design charts for r_u = 0.25 and 0.5 after Jewell [14].

Method	Moment Equilibrium	Force Equilibrium
Ordinary or Fellenius	Yes	No
Bishop's Simplified	Yes	No
Janbu's Simplified	No	Yes
Spencer	Yes	Yes
Morgenstern-Price	Yes	Yes
Janbu Generalized	Yes (by slice)	Yes
Sarma – vertical slices	Yes	Yes

Table 3.
Equations of statics satisfied (after Krahn [50]).

Method	Interslice Normal (E)	Interslice Shear (X)	Inclination of X/E Resultant, and X-E Relationship
Ordinary or Fellenius	No	No	No interslice forces considered
Bishop's Simplified	Yes	No	Horizontal
Janbu's Simplified	Yes	No	Horizontal
Spencer	Yes	Yes	Constant
Morgenstern – Price	Yes	Yes	Variable; user function
Janbu Generalized	Yes	Yes	Applied line of thrust and moment equilibrium of slice
Sarma – vertical slices	Yes	Yes	X = C + E tan ϕ

Table 4.
Interslice forces and relationships (after Krahn [50]).

6.3.1 Koerner's design method

Koerner [51] proposed a method of slices for analysis of geosynthetic reinforced homogeneous slope neglecting interslice forces. Assuming circular arc failure surfaces minimum FS is found by varying the radius and coordinates of the origin of the circle. For slope reinforced with horizontal layers (**Figure 10**) of geosynthetics, FS, is

$$FS = \frac{\sum_{i=1}^{i=n} (Ni \tan\phi + c\Delta li) R + \sum_{j=1}^{j=m} TjYj}{\sum_{i=1}^{i=n} (wi \sin\theta i) R} \tag{2}$$

where W_i = weight of ith slice, θ_i = angle made by tangent to the failure arc at the center of ith slice with the horizontal, $N_i = W_i \cos\theta i$, Δl_i = arc length of ith slice, R = radius of circular curve, c and ϕ- strength parameters, T_j = allowable tensile strength of geosynthetic at j[th] layer, y_j = moment arm for j[th] layer, m = number of geosynthetic layers, n = number of slices. For fine grained soil, the equation for FS simplifies to

$$FS = \frac{cL_{arc} R + \sum_{i=1}^{m} T_i i Y_i}{WX} \tag{3}$$

where W = weight of circular slice and X is the horizontal distance of CG of soil mass from the center of the critical slip circle.

6.3.2 Generalized limit equilibrium method

A generalized limit equilibrium (GLE) formulation was developed by Fredlundand and Krahn [52] and Fredlund et al. [53]. This method encompasses the key elements of all the methodslistedin **Table 2**. The interslice shear forces (Morgenstern and Price [49]) are

$$X = E\lambda f(x) \tag{4}$$

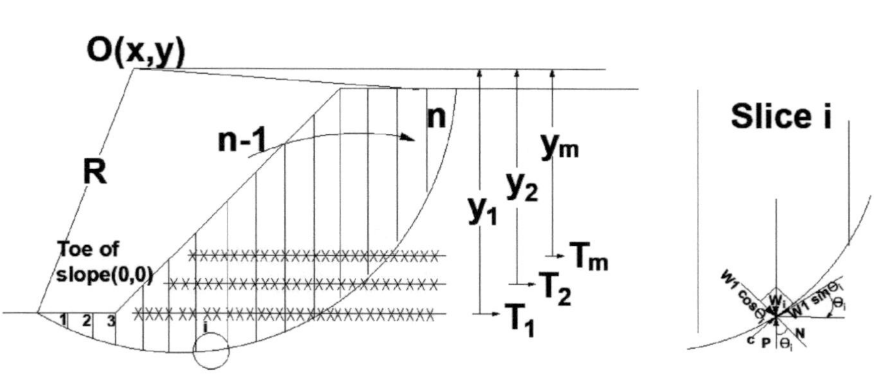

Figure 10.
Circular arc slope stability analysis for geosynthetic reinforced c- ϕ soil (after [44]).

Where f(x) –a function, E and X –the interslice normal force and shear forces respectively and λ- function. GLE Method showing forces on slice and geometrical parameters is shown in **Figure 11**.

The factor of safety, F_m, with respect to moment equilibrium is

$$F_m = \frac{\Sigma\left[C'\beta R + (N - u\beta)R\tan\phi'\right]}{\Sigma Wx - \Sigma Nf \pm \Sigma Dd} \tag{5}$$

The factor of safety, F_f, with respect to horizontal force equilibrium is

$$F_f = \frac{\Sigma\left[C'\beta\cos\alpha + (N - u\beta)\tan\phi'\cos\alpha\right]}{\Sigma N\sin\alpha - D\cos\alpha} \tag{6}$$

where c' –effective cohesion, ϕ'-effective angle of friction, u –pore water pressure, N –normal force on slice base, W–slice weight, D–line load and β, R, x, f, and d – geometric parameters as detailed in **Figure 11**, and α - inclination of slice base. One of the key variables in both the equations is N –the normal force at the base of each slice. N is obtained by the summation of vertical forces as

$$N = \frac{W + (XR - XL) - \dfrac{C'\beta\sin\alpha + u\beta\sin\alpha\tan\phi}{F}}{\cos\alpha + \dfrac{\sin\alpha\tan\phi'}{F}} \tag{7}$$

The base normal force, N, is dependent on the interslice shear forces respectively XL and XR on the left and the right sides of the slice. The reinforcement force is accounted for in the analysis by GLE method.

6.3.3 Finite element computed stress in limit equilibrium

Krahn [45] suggested that normal stress determined from finite element stress analysis can be fed into General Limit Equilibrium Analysis. Thus, Limit Equilibrium and Finite Element method are integrated.

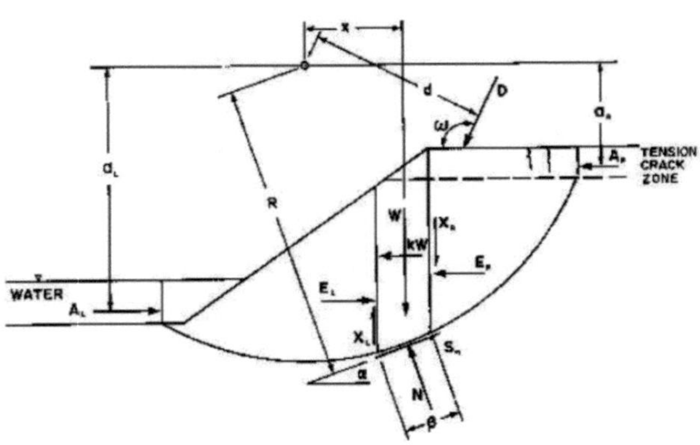

Figure 11.
Forces acting on sliding mass with circular slip surface (after [47]).

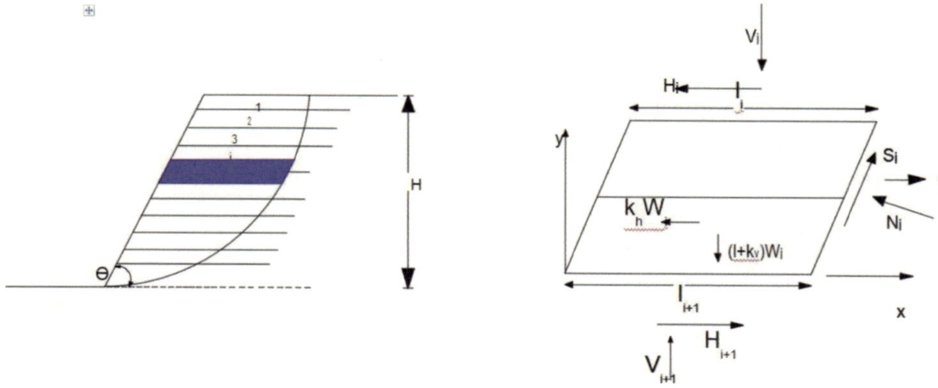

Figure 12.
Horizontal slice method (after Shahgholi and Fakher [39]).

6.3.4 Iterative GLE method for reinforced slope

Song et al. [28] proposed new approach based on LE principle to evaluate stability of reinforced slope. The effect of reinforcement is included as equivalent resisting force acting at slice base and added to GLE method. The corresponding equations are derived based on force equilibrium in the directions normal and parallel to slice base and moment equilibrium at the center of base of slice. The indeterminacy is resolved by assuming half sine function for inclination of inter-slice force in Eq. 4. The method satisfies both the force and the moment equilibrium considerations applicable to arbitrary failure surfaces and is iterative.

6.3.5 Horizontal slice method

Shahgholi and Fakher [39] proposed horizontal slice method (HSM) in which horizontal slices are used in place of vertical ones to analyze the stability of reinforced and unreinforced slopes and walls. The limitation of the vertical slice method for the analysis of reinforced soil of unknown parameters being more than the number of equations available, is resolved by the horizontal slices method. The assumptions of HSM are (i) the vertical stress on an element in the soil mass is equal to the overburden pressure, (ii) the factor of safety (F.S.) is equal to the ratio of the available shear resistance to the mobilized shear stress along the failure surface, (iii) the factor of safety for all the slices is equal and (iv) the failure surface can have any arbitrary shape but does not pass below the toe of the slope or wall. Forces acting on a horizontal slice are shown in **Figure 12**.

With failure wedge divided into N horizontal slices, 4Nunknowns can be determined by 4 N equations and a complete formulation is possible. The formulation is simplified if only vertical equilibrium is considered for individual slices together with overall horizontal equilibrium for the whole wedge, no account being taken of moment equilibrium. In this case the number of equations and the number of unknowns get reduced to 2 N + 1. HSM was extended to design of RE walls considering oblique pull by Reddy et al. [54].

7. Reinforced slope with cohesive backfill

Reinforced Slopes are conventionally constructed with granular fill. However, this has limited the use of reinforced soil structures in locations where such free draining backfill material is not readily available in close vicinity of the sites. Zornberg and Mitchell [55] and Mitchell and Zornberg [56] evaluated the use and performance of reinforced soil structures constructed with poorly draining and/or cohesive backfills. Permeable reinforcements are particularly appropriate for poorly draining backfills as they facilitate dissipation of excess porewater pressures.

7.1 Design methods

Christopher et al. [29] provide design guidance for reinforced soil structures using poorly draining backfills, viz., total stress analysis ignoring the drainage contribution of geocomposite for short term and effective stress analysis considering drainage in the long term. Naughton et al. [30] improved the design method of Christopher et al. [29] and presented single stage effective stress analysis since excess pore pressure gets dissipated fully before construction of subsequent layers. Half meter thickness of each lift is proposed to control short term stability of the slope face. Giroud et al. [32] updated design method of Naughton et al. [30] for reinforced slopes using draining geogrid with focus on an improved determination of the required transmissivity of the same. The method is practical as it makes it possible to optimize the design by adjusting the parameters such as the construction time, time required for pore pressure dissipation, layer thickness and drainage length. Naughton et al. [57] present a parametric study of design parameters of low permeability fill and concluded that for typical compressibility and consolidation parameters vertical spacing of the reinforcement of 0.5 m was adequate.

7.2 Design method - Giroud et al.

Giroud et al. [31, 32] presented a method for reinforced slopes using draining geogrid. Transmissivity and length of draining path are important for geogrids from the design point. Typical values of the parameters are: ϕ ‘of 20° to 30°, drained cohesion, c’ of 0–20 kPa, coefficient of consolidation, c_v of 0.1–100 m^2/year, and coefficient of compressibility, m_v of 0.01–5 m^2/MN. Long term hydraulic transmissivity, θ_a, of draining geogrid is obtained by applying a set of reduction factors to the laboratory measured value. Reduction factors account for creep, particulate, chemical and biological clogging of drainage channels.

Geometry of reinforced fill is important from design point of view. A drain located at the back of reinforced zone (**Figure 13**) is generally used to prevent groundwater from flowing into the reinforced zone and to halve the drainage length

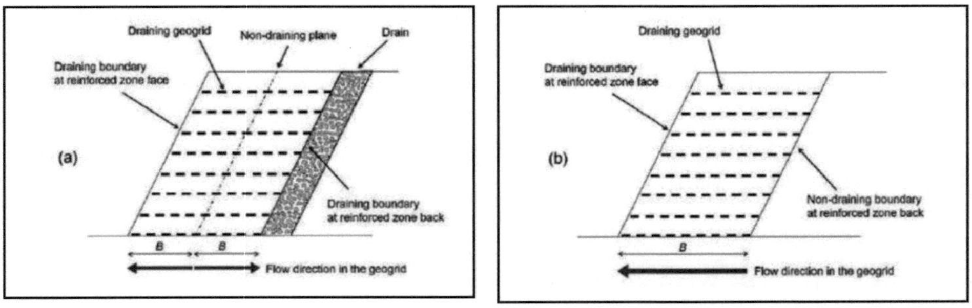

Figure 13.
Geometry of reinforced slope with (a) and without drain (b) (after Naughton et al. [32]).

in the draining geogrid. Design has two main components, viz., (i) determination of required transmissivity of draining geogrid and required time for rapid dissipation of porewater pressure and (ii) determination of stability and settlement of slope as in conventional practice.

The maximum porewater pressure in the drainage channel should not be too high nor too low. If the porewater pressure in the drainage channel is too high, vertical flow of water from the fill to the drainage channel will be slowed down, whereas, if the water pressure in the drainage channel is too low, flow of water will be too slow. Solution to this complex problem is presented by Giroud [58]. The rate of maximum vertical flow rate from fill to drainage channel depends on time factor ($T_0 = 4C_v t_0/H^2$) and occurs at the end of construction, t_0, of fill layer of thickness, H, overlying draining geogrid.

The following equations for the required transmissivity, θ_{req}, are derived by Giroud [58], assuming that the maximum water pressure in the drainage channel, u_{max}, is 10% of the overburden stress, consistent with a degree of consolidation of 90% in the fill.

$$\theta_{req} = 10\mathrm{B}^2\mathrm{k} / \mathrm{H}\sqrt{T_0} = 5\mathrm{B}^2\mathrm{k} / \sqrt{C_v t_0} \text{ if } 1\mathrm{x}10-6 \le T_0 \le 1 \tag{8}$$

and

$$\theta_{req} = 10\mathrm{B}^2\mathrm{k} / \mathrm{HT0} = 5\mathrm{B}^2\mathrm{k} / 2\mathrm{C}_v\mathrm{t}_0 \text{ if } 1\mathrm{x}10-6 \le T_0 \le 1 \tag{9}$$

The required transmissivity, θ_{req}, must be less than the allowable transmissivity, θ_a. Parameter B in the above equation is dependent on length of draining geogrid. In case of draining boundary at the back of the reinforced zone (**Figure 13**) drainage path length is equal to half and in case of non-draining boundary it is equal to full length of reinforcement. The required transmissivity depends on hydraulic gradient in the drainage channel. An approximate value of hydraulic gradient in the drainage channel is given by

$$\mathrm{i}_{avg} = \mathrm{u}_{max} / \gamma_w \mathrm{B} \tag{10}$$

where u_{max} is maximum allowable water pressure in drainage channel and is generally taken as 10% of overburden stress. The time required for pore water pressure dissipation is estimated as time required to reach 90% consolidation and is calculated as under

$$t_{90} = \frac{T_{90}H^2}{4C_v} \tag{11}$$

where T_{90} is time factor for 90% consolidation. The parameter B is equal to the length of reinforcement and is obtained as part of stability analysis (Jewell [14]). Alternatively, the geogrid length is selected arbitrarily and stability is checked separately.

7.2.1 Parametric study

Naughton et al. [32] conducted parametric studies to study time for pore pressure dissipation, and required transmissivity of the geogrid. From practical consideration, construction of one layer per day implies about 24 hours for porewater pressure dissipation. Studies reveal that this is achievable unless c_v is less than 30 m^2/year and vertical spacing is more than 0.5 m. For c_v values greater than 50 m^2/year and thickness of fill not exceeding 0.5 m, the dissipation time required reduces to less than 12 hours (**Figure 14**).

For soils with c_v up to 50 m^2/year, the unfactored required transmissivity in the draining geogrid is less than 1.2×10^{-6} m^2/s. As the drainage characteristic of the soil improves, i.e., $c_v > 75$ m^2/year, the required transmissivity increases rapidly by orders of magnitude (**Figure 15**). The required transmissivity also depends on the vertical spacing of the draining geogrid.

Smaller vertical spacings and longer reinforcement lengths require larger transmissivity in the draining geogrid. Reinforcement spacing of 0.5 m optimizes the time for dissipation of pore pressures while, at the same time, requiring realistic and achievable transmissivity in the draining geogrid.

At this spacing a draining geogrid will dissipate pore pressures over the full range of likely values encountered in low permeability fills within 24 hours, with the further advantage that the required transmissivity is independent of reinforcement length. With above spacing, reinforced slope with poorly draining backfill can be analyzed as a normal slope for both internal and external stability for normal ranges of soil parameters.

7.3 Semi-analytical design method for cohesive backfill reinforced slope

Abd and Utili [33] developed a semi-analytical method for uniform slope with c-ϕ soil using Limit Analysis (LA). The method provides the amount of reinforcement needed as a function of cohesion, tensile strength, angle of shearing resistance and slope inclination. Climate induced crack and cracks that form due to slope collapse are accounted for in this method. Both soil and reinforcement are assumed to be rigid-plastic and follow normality rule i.e., associated plastic flow.

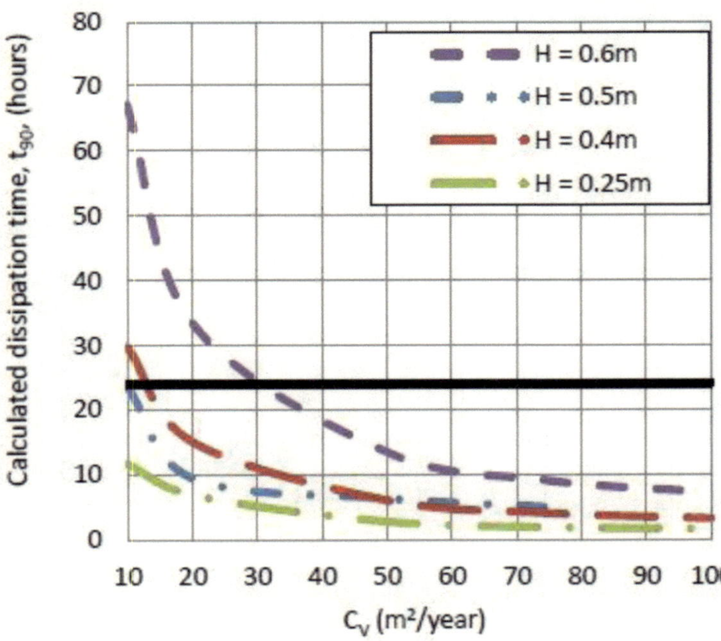

Figure 14.
Variation of dissipation time (t_{90}) with coefficient of consolidation (c_v) for reinforcement length equal to height of slope (after Naughton et al. [32]).

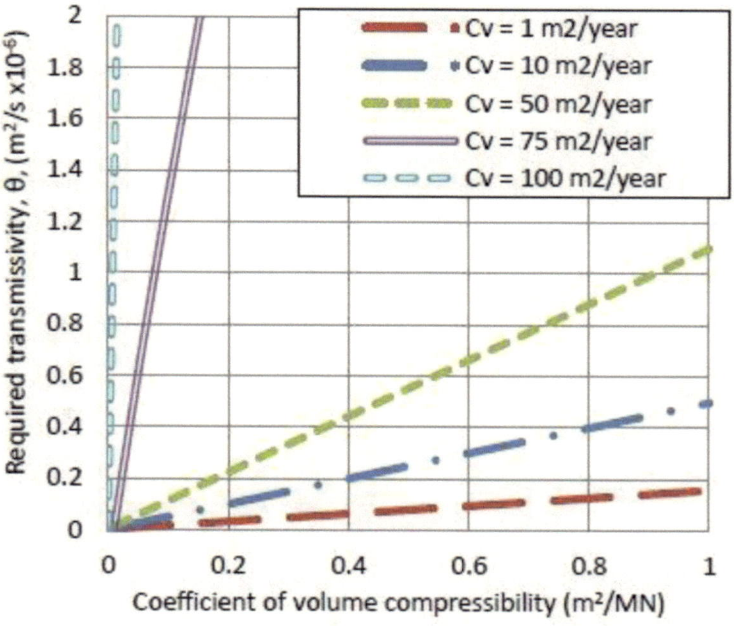

Figure 15.
Variation of required transmissivity of the geogrids, θ, with coefficient of volume compressibility, m_v, for a range of coefficients of consolidation, reinforcement length = 1.0H (after Naughton et al. [32]).

7.3.1 Design

Traction free slopes with slope angle of 40–90° reinforced with geosynthetic layers are considered. Reinforcement is equally spaced throughout (uniform distribution - UD) or at a spacing decreasing from top to bottom of slope (Linearly Increasing Distribution - LID) (**Figure 16**).

Average strength of reinforcement, K_t, for UD case is

$$K_t = nT / H \tag{12}$$

where n- the number of reinforcement layers, T - the strength of a single layer at yield and H - the slope height.

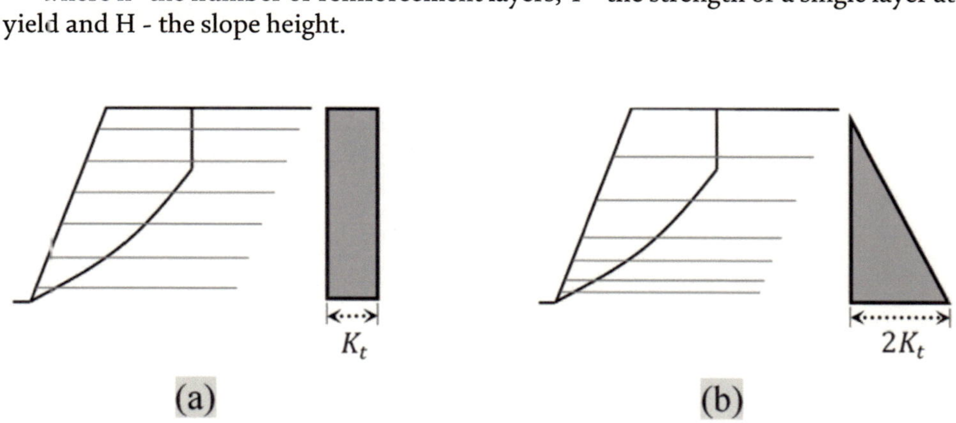

Figure 16.
Geosynthetic-reinforcement layouts: (a) uniform and (b) linearly increasing distribution with depth (after Abd. and Utili [57]).

For LID case local reinforcement strength, K, is

$$K = 2K_t\left(H - y\right)/H \tag{13}$$

where y - the vertical coordinate from the slope toe. Maximum depth of crack is limited from the requirement that remaining slope remains stable. Upper bound maximum crack depth to be adopted [59] is

$$h_{max} = \frac{3.83c\tan\left(\frac{ð}{4}+\frac{ö}{2}\right)}{y} \tag{14}$$

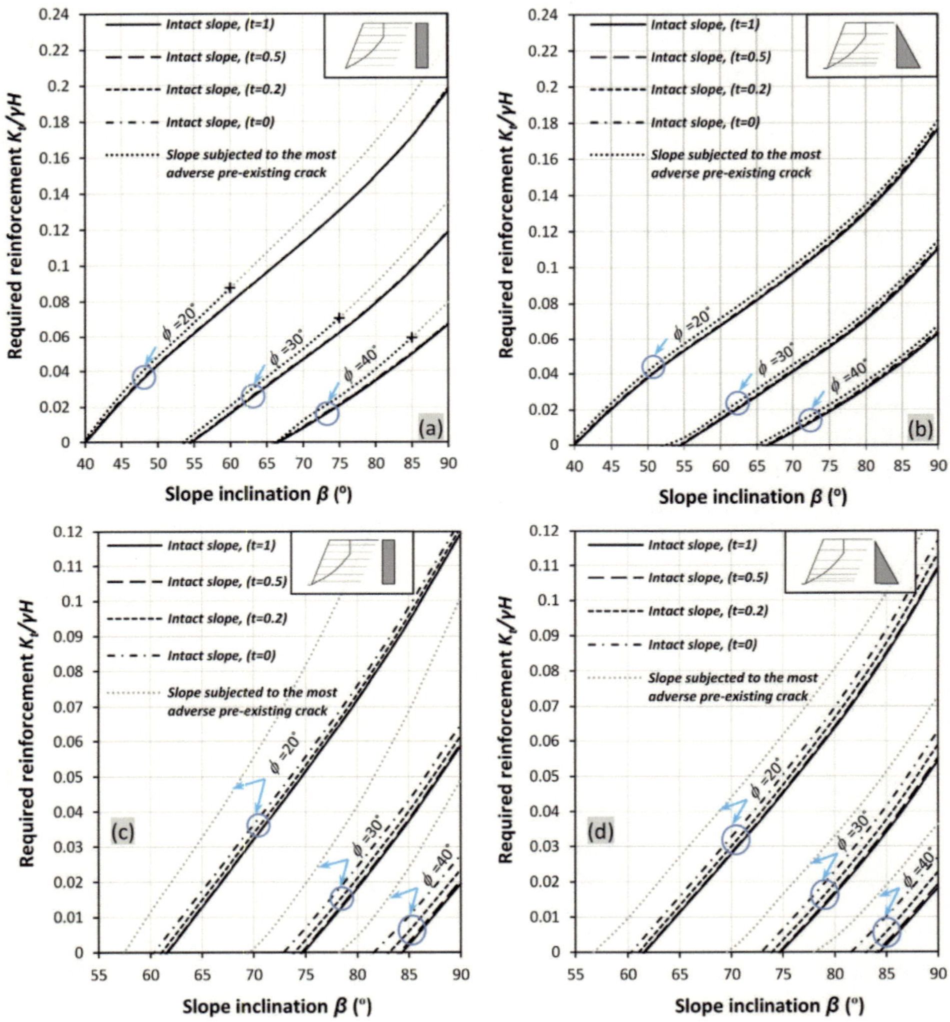

Figure 17.
Design charts for intact slopes not subject to crack formation (t = 1), intact slopes subject to crack formation (limited tensile strength of t =0.5, t = 0.2 & t = 0) and cracked slopes. (a) & (b) are for c/γH = 0.05 while (c) & (d) are for c/γH = 0.1. (after Abd and Utili [33]).

7.3.2 Required reinforcement

Design charts (**Figure 17**) provide the reinforcement strength and embedment length for uniform and linearly increasing reinforcement distributions for different slope angles β and ϕ for specified value of c/γH.

In **Figure 17**, 't' is dimensionless coefficient representing soil tensile strength and is defined as ratio of ground tensile strength to be measured experimentally over maximum unconfined tensile strength consistent with Mohr-Coulomb criteria. Considering the case of intact slopes, it can be observed that for relatively low values of cohesion, c/ϒH = 0.05, the tensile strength, t, has negligible effect on the required reinforcement force. But for higher values of cohesion (c/ϒH = 0.1), t becomes important. Above charts are for fully drained slopes.

Using **Figure 17**, $K_t/\gamma H$ can be determined for given slope angle, β, and angle of shearing resistance, ϕ, of soil. Tensile strength of reinforcement is calculated using Eq. 10 for given number of layers. Influence of porewater pressure on required amount of reinforcement is analyzed using r_u method [60]. A uniform value of r_u is assumed throughout the slope and effective stress analysis carried out. **Figure 18** provides $K_t/\gamma H$ for slope inclinations of 40^0 to 90°, r_u = 0, 0.25 and 0.5 for UD and LID cases.

Gray and black lines in **Figure 17** indicate respectively active and inactive constraint of maximum crack depth. The mark + signals the boundary between the two.

Gray and black lines in **Figure 18** indicate respectively active and inactive constraint of maximum crack depth. The mark + signals the boundary between the two (after Abd and Utili [33]).

A combined failure mode consisting of pullout in some layers and rupture (tensile failure) in others, also needs to be considered to calculate the minimum length of the reinforcement layers. **Figure 19** provides L_r/H as a function of slope angle, β for ϕ of 20^0 for this case.

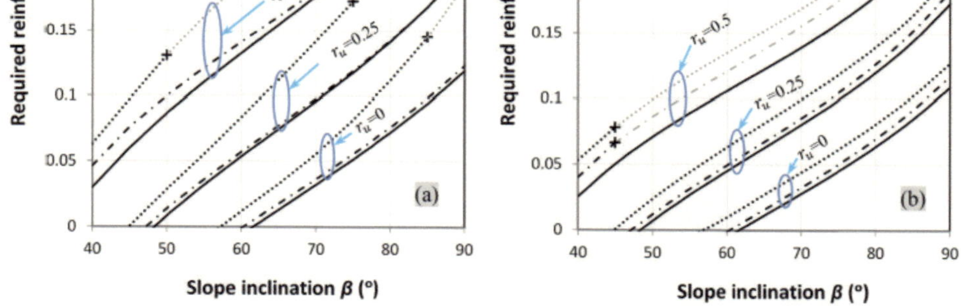

Figure 18.
Design charts for the required reinforcement for intact and cracked slopes (with ϕ = 20° and c/γH =0.1) (a) UD of reinforcement; and (b) LID.

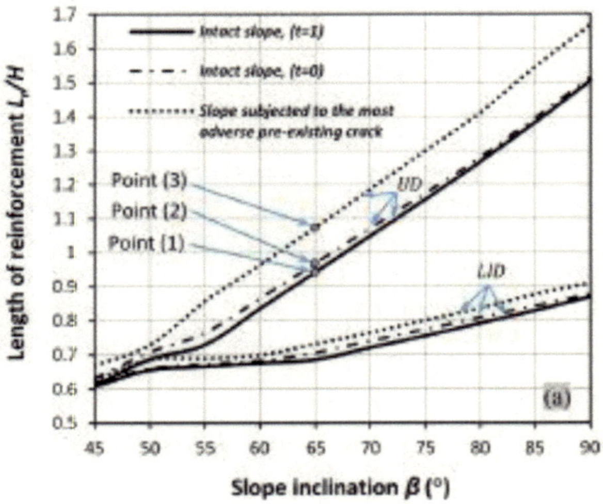

Figure 19.
Length of reinforcement for a slope with ϕ = 20°, $c/\gamma H$ = 0.05 *and* r_u = 0.

8. Slope - reinforcement interaction and length optimization

The inclusion of geosynthetic reinforcement in soil slope for improving stability leads to change in behavior of reinforced soil mass due to induced stresses as compared to that in unreinforced slope so far as critical slip circle is concerned. Hence there is a need to understand slope-reinforcement interaction behavior. Jha et al. [61–64] optimized reinforcement length form face of slope and identified and quantified slope-reinforcement interaction.

To study soil reinforcement interaction and length optimization reinforced embankment slope, on a competent soil, 6.0 m high with side slope of 1.5H:1 V is considered (**Figure 20**).

The embankment and foundation soil have cohesion, c, of 5 kPa, unit weight, γ, of 18 kN/m^3 and angle of shearing resistance, ϕ, of 230. The geotextile reinforcement used has adhesion, ca, of 3 kPa, angle of interface friction between soil and reinforcement, δ, of 17° and ultimate tensile strength, T_{ult}, of 200 kN/m.

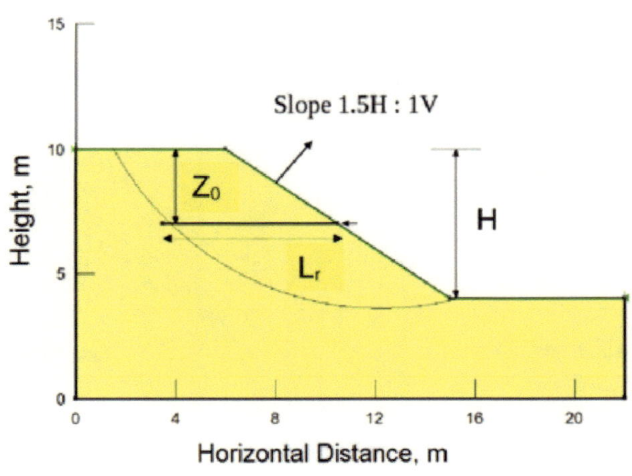

Figure 20.
Schematic diagram - reinforced embankment slope.

Analysis of unreinforced embankment of 6 m leads to FS_{min} of 1.22 less than the required value of 1.3 and hence needs to be reinforced with geosynthetic reinforcement to get preferably long term FS_{min} of 1.5.

8.1 Reinforcement length optimization: non-face end

Effect of varying the length, L_r, of geosynthetic placed at depth, Z_0 = 3.0 m in 6.0 m high embankment is studied by curtailing it from the non-slope face to get FS_{min} in the range of 1.50 to 1.60. The length, L_r, of the reinforcement to intercept the failure surface at 3.0 m depth was varied from 8.0 m with FS_{min} of 1.6 (**Figure 21**).

Circles ABC and DEF are the critical slip circles of the unreinforced and the reinforced slopes. One of the effects of inclusion of reinforcement in embankment soil is to shift the critical slip circle from ABC to DEF, that is from shallower to deep inside the slope. This shift of the critical circle therefore increases the factor of safety by involving larger slide mass. PQ is reinforcement of length L_r. The length of reinforcement L_r has two components: QE = effective length, L_e, in the stable zone and EP – the length, L_f in the unstable zone. L_f is further divided into lengths L_{f1} (EB), the length in the failure zone between the critical slip circles of the reinforced and the unreinforced slopes and length, L_{f2}, between the critical slip circle of unreinforced slope and slope face (BP) as shown in **Figure 21**. The effect of varying L_r with right end fixed at point P and left end (Q) curtailed inwards successively, leads to reduction in mobilized force in the reinforcement (F_r) from 35.8 kN/m to 19.6 kN/m corresponding to reduction in L_r from 8.0 to 7.3 m. FS_{min} reduces from 1.60 to 1.51. Factor of safety and the load/resistance mobilized in the reinforcement decrease with reducing length of reinforcement as is to be expected. FS_{min} falls below 1.50 on reducing the length further to 7.0 m.

8.2 Length optimization from face end

The length, $L_f = (L_r - L_e)$ is much larger than L_e, the effective length of reinforcement contributing to increase in the stabilizing moment/force. The required pullout force in the reinforcement in the stable zone should equal to that mobilized by the corresponding length of the reinforcement in the unstable zone for equilibrium. It would serve no useful purpose if the length of the reinforcement in the unstable

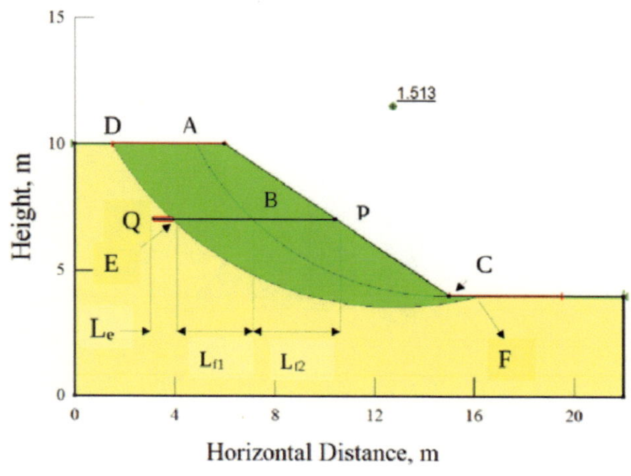

Figure 21.
Critical slip circle for Z_o = 3.0 m, FS_{min} = 1.51, L_r = 7.27 m.

zone is more than that required for generating the required stabilizing force. Hence minimizing $L_f = (L_r - L_e)$ by moving point P inside the soil mass and away from the slope face by curtailing length of reinforcement but still maintaining FS_{min} above 1.50 can lead to economy. Accordingly, for reinforced slope of **Figure 21**. L_r has been curtailed from the face end of the slope. As point P is moved inside gradually by reducing L_r, the critical circle continues to be DEF or close to it (**Figure 22**), i.e., practically with no shift of the critical circle.

The minimum length, L_r which provides $FS_{min} = 1.51$ is obtained as 5.08 m (**Figure 22**). Thus about 30% reduction in length of reinforcement is achieved without sacrificing the stability of the embankment slope as FS_{min} is still the required value of 1.5. Hence length optimization from face end leads to saving of reinforcement length. The circle, ABC, is not the critical for the reinforced slope and thus not acceptable as the critical circle with consideration of reinforcement is different from that of unreinforced case.

8.3 Slope-reinforcement interaction

Slope as in **Figure 21** has been analyzed further for the critical slip circle DEF of reinforced slope but without considering the effect of reinforcement to get FS of 1.41 (**Figure 23**).

The summary of results for various depth of reinforcement from top of embankment is summarized in **Table 5**.

The contribution of reinforcement in enhancing the stability of a slope is observed to be twofold: (i) shifting of the critical slip circle deeper in to the slope involving larger slide mass or forward involving smaller slide mass and thus enhancing the factor of safety of the slope and (ii) due to contribution of reinforcement to stabilizing force/moment. FS_{min} of 1.22 for unreinforced case increases to 1.41 due to shifting of the critical circle to DEF an increase of 15.6%. Secondly the contribution of reinforcement to stabilizing moment/force leads to a further increase in factor of safety from 1.41 to 1.51, a contribution of about 8.2%.

8.4 Summary of slope -reinforcement interactions

Slope - reinforcement interaction analysis is summarized as follows: (i) The critical slip circle for the slope with reinforcement shifts inward and is very different

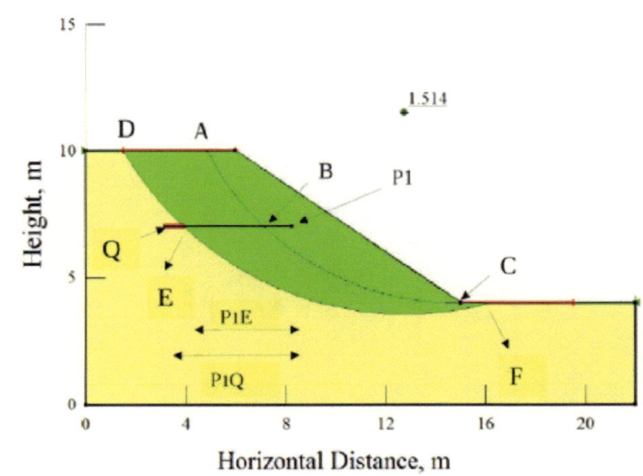

Figure 22.
Critical slip circle for slope with Zo = 3.0 m, Lr = 5.08 m and FSmin = 1.51.

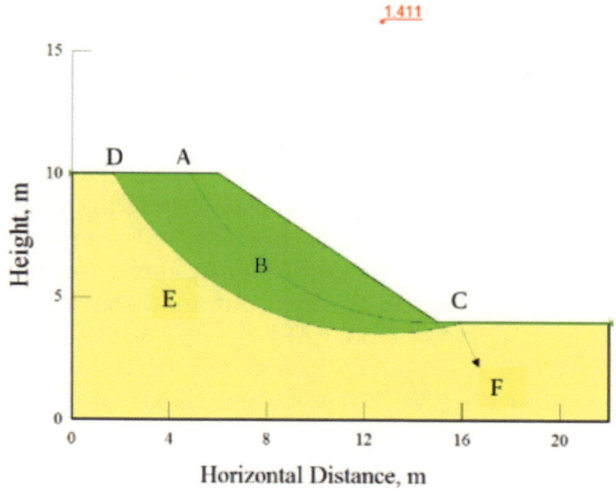

Figure 23.
Slope stability with critical slip circle DEF but without considering the effect of reinforcement.

	FS				L_r, m
Z_0, m	I	II	III	IV	
3.0	1.22	1.51	1.80	1.41	5.08
4.0	1.22	1.51	1.86	1.48	5.26
5.0	1.22	1.51	1.92	1.46	6.04

I: *FS_{min} for unreinforced slope with critical circle ABC;* **II**: *FS_{min} for reinforced slope with critical circle DEF;* **III**: *FS for reinforced slope analyzed for circle ABC of unreinforced slope and* **IV**: *Reinforced slope analyzed for critical slip circle DEF but without considering the effect of reinforcement.*

Table 5.
Factors of safety and lengths of Geosynthetics.

from that for the unreinforced slope; (ii) The increase in factor of safety is because of the shift of the critical slip circle deep in the slope and involving larger sliding mass. This results from the fact that the slip circle is deeper in to the soil and away from the critical circle corresponding to that for unreinforced embankment soil; As a consequence, the reinforcement force generated becomes much smaller than that estimated based on the length corresponding to that estimated with respect to slip circle for the unreinforced slope; (iii) The effect of providing reinforcement in the slope is two-fold, viz., shifting of critical circle inside of the embankment involving larger slide mass and by increase in stabilizing force/moment due to bond resistance mobilized in the reinforcement; and (iv) It is possible to achieve about 20 to 30% shorter length of the reinforcement without endangering the stability of the embankment slope. The most significant finding of this study is that the reinforcement can be provided from inside and not necessarily from the face of the embankment.

9. Software for analysis of reinforced slope

The objective in designing geosynthetic reinforced soil slope is to determine the required long-term strength and layout of the reinforcement apart from finding the critical failure surface. The layout and strength are interrelated rendering many possible solutions with the same level of stability but not necessarily having the same economics. Software facilitates the designer to reach an optimal solution apart from locating critical failure surface by using search techniques and by repeatedly performing the stability calculations for different failure surfaces.

Pockoski and Duncan [65] compared several Limit Equilibrium based software available based on features of program, ease of use, range of applicability, accuracy, and efficiency. These programs were rated as well considering accuracy of results, computation time, learning curve, time to enter data and complete an analysis, ease of reinforced slope design, ease of unreinforced data entry, time required to make graphical output Report -Ready and quality of output. Different software included for comparison are: UTEXAS4 & TEXGRAF4, SLOPE/W, SLIDE, XSTABL, WINSTABL, RSS, SNAIL, GoldNail. UTEXAS4 is a precise analysis tool but does not have graphic user interface. TEXGRAF4 is second part of UTEXAS software package and displays information and results of the UTEXAS4 search and generates file for use in CAD software. SLOPE/W having a graphic user interface is user friendly and versatile. Reinforcement inclusion in to the analysis is also graphical. SLOPE/W has Monte Carlo based probabilistic stability analysis option where by soil, porewater pressure and seismic coefficient can be entered with standard deviation. SLIDE is Windows based slope stability program and can search for a critical circular, non- circular or composite slip surfaces. XSTABL is an interactive program which can search critical circular surface. WINSTABL is windows-based program whereby geosynthetic reinforcement, anisotropic soil, seismic loads, etc. can be considered in the analysis. RSS is an interactive program and is capable of exhaustive search performed on reinforced slope. Circular, bilinear, bottom third and top third failure surfaces are considered. SNAIL is window based free software and permits use of seven different types of soils and uses force equilibrium on two and three-part wedge analysis. GoldNail is a very powerful and design program that is primarily meant for soil nail wall analysis.

SSAP release 5.0 (2020) software (https://www.ssap.eu) is a versatile free software and uses advanced limit equilibrium method and FE-LEM combination to get the critical Factor of safety. ReSlope is an interactive, design-oriented, program for geosynthetic-reinforced slopes. For a given problem including geosynthetic strength, reduction factors, and design safety factors, ReSlope produces the optimal layout (i.e., length and spacing) of reinforcement layers. ReSlope was specifically developed for geosynthetics. SVSLOPE, GeoStru, Oasys, ReActive, Secuslope are other software available for reinforced slope stability analysis.

10. Case studies and lessons learnt

Liu et al. [66] studied failure of a four-tiered geogrid reinforced slope of a road embankment of height varying from 10 m to 40 m over a length of 430 m to assess mechanism and causes contributing to these failures (**Figures 24** and **25**). A flat natural slope of 28° was converted to a steep geogrid reinforced slope of 0.5 H:1 V (63°) slope for constructing an approach road.

The first slope failure occurred during the construction phase itself soon after the rainy season as rainwater seeped into permeable laterite gravel layer underlain by an impermeable clay layer. The interface of laterite gravel and clay created a detrimental bedding plane whose shear strength was reduced due to infiltration of water. The slide got initiated along the interface when toe was excavated to construct the reinforced zone. The second failure occurred due to very strong earthquake. The overstress initiated near the vicinity of the clay layer extended into the retained natural slope to form a massive slide. The third failure occurred following

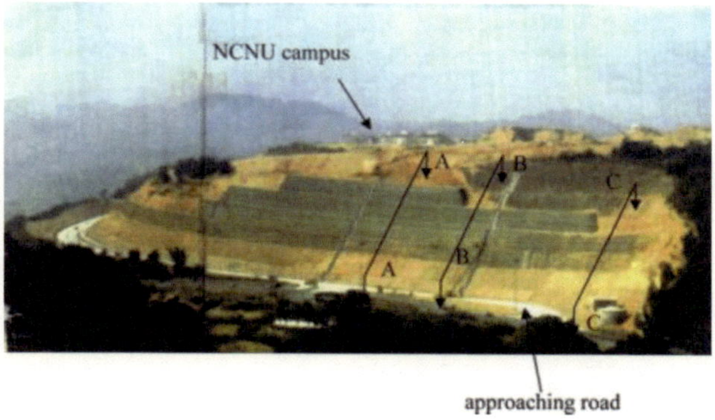

Figure 24.
Approach road (after [66]).

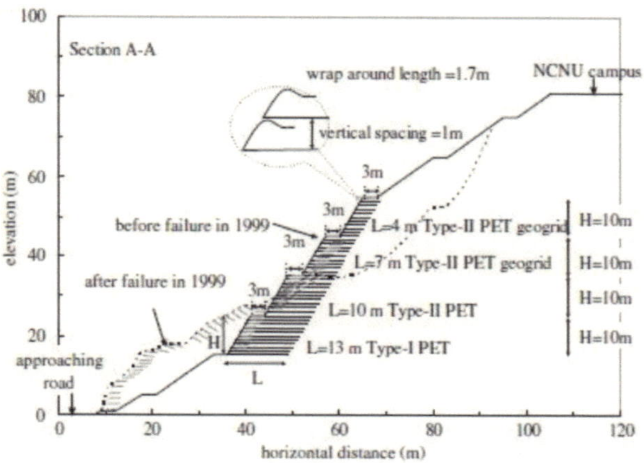

Figure 25.
Reinforced slope with failure details (after [66]).

abundant rainfall during a heavy rainstorm that infiltrated into the reinforced slope as no sub-drainage system was provided. The infiltration that was obstructed by the impermeable clay and fine contents in the backfills generated significant water pressure, inducing the slope to fail behind the reinforced zone.

The interface between laterite gravel and clay is an embedded weak plane, which when saturated softened. In addition, because of its low permeability, it became a barrier to the infiltration. Site investigation failed to find the existence of this clay layer, because the reinforced slope was thought to be subsidiary to campus buildings, and no investigation efforts made specific to the reinforced structures. The succeeding design and construction did not appropriately correspond to this clay layer even when it was observed during construction. The results showed the impact of the clay layer on the slope stability to be very critical. The current practice of considering cohesion needs to be re-evaluated as this apparent cohesion may get reduced to even zero with increasing saturation. The lack of a drainage system was another significant cause for failure.

Yang et al. [67] investigated 26 m high, four-tier geogrid reinforced slope (**Figure 26**), backfilled with low plasticity silty clay that contains more than 60% of fines (marginal backfill). Prior to the completion of construction, tension cracks were discovered along with slope settlement at the top of the slope. The tension cracks and slope settlement were caused by a series of heavy rainfall. The slope

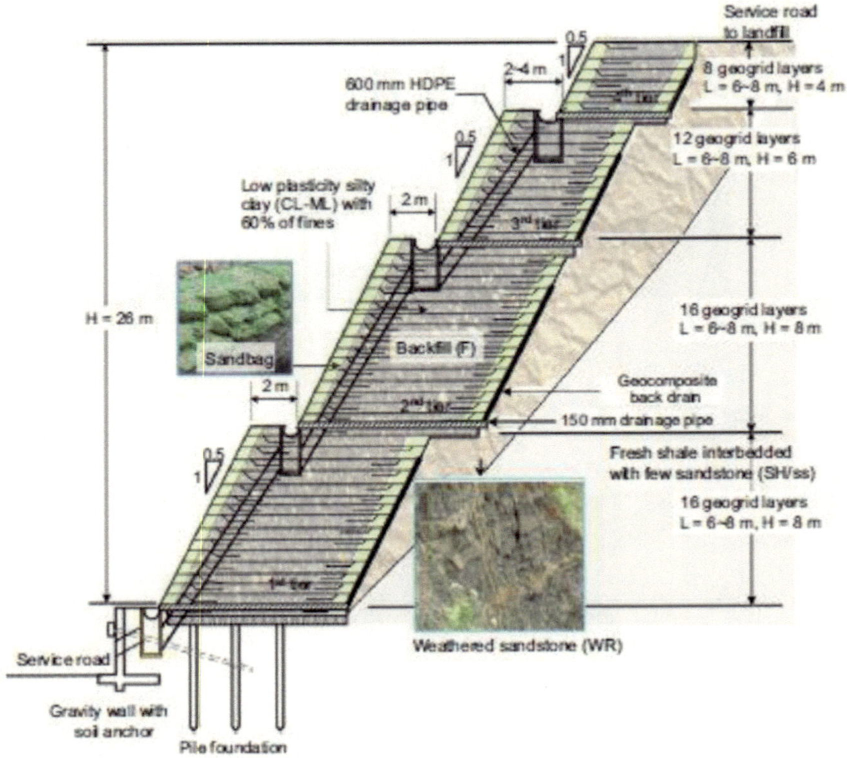

Figure 26.
Layout and Design of Multitier Geosynthetic Reinforced Slope (after [67]).

settlement was repaired by placing additional backfill on the top of the slope to compensate for the settlement that had occurred.

During the next rainy season, the slope was subjected to a significant amount of rainfall of 187 mm in May, 350 mm in June, 243 mm in July, and 563 mm in August. During this period, tension cracks and slope settlement got regenerated and gradually developed as the rainfall continued. **Figure 27** displays subsequent development of the tension cracks and slope settlement with time.

Factors that caused the slope failure from the forensic investigation are: (i) The use of marginal soil (over 60% of fines) as the backfill without provision of drainage. (ii) The original design and site investigation overlooked the existence of the weathered and fractured rock layer, which has shear strength less than that of an intact rock. (iii) Tension cracks and slope settlement developed at the top surface of the slope allowed rainwater to pond on the top and to infiltrate into the reinforced zone. (iv)The drainage system may have malfunctioned as joints were poorly and loosely connected and likely got dislocated due to the excessive deformation.

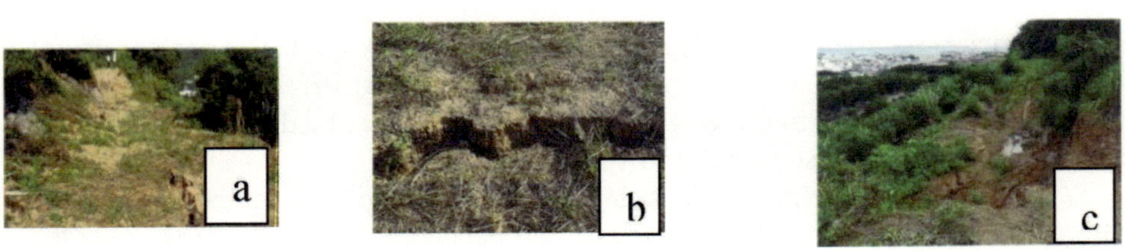

Figure 27.
Tension crack and settlement of slope and its failure after rains of 2012(a) tension crack; (b) onset of settlement; (c) excessive settlement over 1 m (after [67]).

Lessons learnt from these case histories are: (i) Detailed site investigation should be carried out to asses presence of weak layer, soil weathered rock, etc., (ii) Design cohesive backfill slope for drainage and with provision of draining geosynthetics, (iii) Install drainage systems appropriately and (iv) Design RE slope for global stability.

However, it is pertinent to mention that marginal soil can be used with draining geogrids as detailed in Section 7.2 with adequate drainage capacity.

11. Conclusions

Steep slope embankment is a necessity for development of rail, road and other infrastructure projects. Safety of embankment slopes is of utmost importance which requires proper site investigation, analysis and design. Limit Equilibrium, Limit Analysis, Slip Line & Finite Element Methods are design methods for RE Slopes. Limit equilibrium method is most commonly used for design including the effect of reinforcement. Jewell's design method [14] for geosynthetic reinforced steep slope soil with granular soil is most commonly used method. Song et al. [28] proposed new approach based on LE principle to evaluate stability of reinforced slope. Slopes with cohesive backfill have been constructed due to limited availability of granular material near project site. Proper design of RE slope using cohesive backfill considering the transmissivity of draining geogrid is important. The method suggested by Giroud et al. [32] for draining geogrid reinforced cohesive back fill slope with 0.5 m/day is practical method as it takes care of pore water pressure dissipation for most common soil parameters. Abd &Utili [33] developed a semi-analytical method for uniform slope with c-ϕ soil using Limit Analysis (LA). The method provides the amount of reinforcement needed as a function of cohesion, tensile strength, angle of shearing resistance and slope inclination. The reinforcement length optimization from face end leads to economy in reinforcement length of the order of 20–30% without affecting factor of safety [61]. There is an interaction between slope and reinforcement [62, 64]. Inclusion of reinforcement in embankment slope results in to shifting of critical slip circle deep inside slope involving larger slide mass thus increasing factor of safety. Reinforcement also provides stabilizing moment/force. Investigation of failed slopes indicate that detailed geotechnical investigation of site to assess presence of weak layer, provision of drainage by way of draining geosynthetics in case of cohesive backfill, installation of drainage systems to capture rain and subsurface water and global stability of reinforced earth slopes are very critical for stability and performance of reinforced slope.

Acknowledgements

I express my sincere thanks to Prof M.R. Madhav my mentor and Co-author of this chapter whose continued encouragement and support made it possible to complete the chapter and bring it to the current format. I am also thankful to my wife Abha who always supported me in this endeavor. I thank Ms. B. Geeta Sahithi my office colleague for helping me for improved figures out of her personal time.

Author details

Akshay Kumar Jha[1*] and Madhav Madhira[2]

1 Indian Railways, Hyderabad, India

2 Department of Civil Engineering, I.I.T. Hyderabad and JNTU, Hyderabad, India

*Address all correspondence to: akshayghunru@gmail.com

References

[1] Petley DN, Dunning SA, Rosser NJ (2005). The analysis of global landslide risk throw the creation of a database of worldwide landslide fatalities. In: Hungr O, Fell R, Couture R, Eberhardt E (eds) Landslide risk management. CRC Press, London, p. 367-373.

[2] Nadim F, Kjekstad O (2009). Assessment of Global High- Risk Landslide Disaster Hotspots. In: Sassa K, Canuti P (eds) Landslide- Disaster Risk Reduction. Springer, Berlin, p. 213-221.

[3] Simac, M.R. (1992). Reinforced slopes: a proven geotechnical innovation. Geotechnical Fabrics Report, p. 13-25.

[4] Zeynep, D. and Tezcan, S. (1992). Cost analysis of Reinforced soil walls. Geotextiles and Geomembranes 11. p. 29-43.

[5] Lostumbo, J. M. (2010). The Yeager Airport Runway Extension: Tallest 1H: 1V Slope in US Case Study Advances in Analysis, Modeling and Design. GSP No.199 ASCE, GeoFlorida, p. 2502-2510.

[6] Devata, M. S (1985). Geogrid reinforced earth embankments with steep side slopes. Polymer Grid Reinforcement, Thomas Telford Limited, London, p. 82-87.

[7] Gharpure A., Kumar S.& Scotto M. (2012). Composite soil reinforcement system for retention of very high and steep fills –A case study. Proceedings Vol 5. Topic: soil improvement and reinforcement. 5th European Geosynthetics Congress. Valencia 2012. P. 346-352.

[8] Jewell, R.A., Paine, N., and Wood, R.I. (1984). Design Methods for Steep Reinforced Embankments. Proc. of the Polymer Grid Reinforcement Conf., London, p. 70-81.

[9] Bonaparte, R., Holtz, R.D. and Giroud, J.P. (1987). Soil reinforcement design using geotextile and geogrids. Geotextile Testing and Design Engineer, ASTM STP 952, J. E. Fluet, Jr., Ed., American Society for Testing Materials, Philadelphia, p. 69-116.

[10] Verduin, J.R. and Holtz, R.D. (1989). Geosynthetically reinforced slopes: A new Procedure. Proc. Geosynthetics'89. San Diego, IFAI, p. 279-290

[11] Leshchinsky D, Reinschmidt AJ (1985). Stability of membrane reinforced slopes. J Geotech Eng. ASCE 111(11). p. 1285-1300

[12] Leshchinsky, D. and Boedeker, R.H. (1989). Geosynthetic reinforced soil structures. J. of Geotechnical Engineering. 115(10). P. 1459-1478.

[13] Schneider, H.R and Holtz, R.D. (1986). Design of Slopes reinforced with geotextiles and geogrids. Geotextiles and Geomembranes. 3. P. 29-51

[14] Jewell, R.A. (1991). Application of the Revised Design Charts for Steep Slopes. Geotextiles and Geomembranes. 10(1091). p. 203-233.

[15] Leshchinsky, D. (1992). Issues in Geosynthetic Reinforced Soil. Earth Reinforcement Practice. Proc. of the International Symposium on Earth Reinforcement Practice. Kyushu University, Fukuoka, Japan, Vol. 2, p. 871-897.

[16] Leshchinsky, D., Ling, H. and Hanks, G. (1995). Unified design approach to geosynthetic reinforced slope and segmental walls. Geosynthetics Int.2(5). p. 845-881

[17] Zhao A (1996) Limit Analysis of geosynthetic reinforced slopes. Geosynth. Int. 3(6). p. 721-740

[18] Michalowski, R. L. (1997). Stability of uniformly reinforced slopes. J. of Geotechnical and Geoenvironmental Engineering, 23(6). p. 546-556.

[19] Shiwakoti, D.R., Pradhan, T.B.S. and Leshchinsky, D. (1998). Performance of geosynthetic - reinforced soil structures at limit equilibrium state. Geosynthetics International.,5(6), p. 555-587.

[20] Baker, R. and Klein, Y. (2004a). An integrated limiting equilibrium approach for design of reinforced soil retaining structures: Part I – Formulation. Geotextiles and Geomembranes, 22(3), p. 119-150.

[21] Baker, R. and Klein, Y. (2004b). An integrated limiting equilibrium approach for design of reinforced soil retaining structures: Part II – Design examples. Geotextiles and Geomembranes, 22(3), p. 151-177.

[22] Han, J. and Leshchinsky, D. (2006), "General analytical framework for design of flexible reinforced earth structures", J. of Geotechnical and Geoenvironmental Engrg. ASCE, 132, 1427-1435.

[23] Leshchinsky, D., Zhu, F. and Meehan, C.L. (2010). Required unfactored strength of geosynthetic in reinforced earth structures. J. of Geotechnical and Geoenvironmental Engineering, 136(2), p. 281-289.

[24] Leshchinsky, D., Kang, B., Han, J. & Ling, H. (2012). Framework of Limit State Design of Geosynthetic Reinforced Walls & Slopes. Transp. Infrastructure Geotech., 1, p. 129-164.

[25] Leshchinsky, B. and Ambauen, S. (2015) Limit equilibrium and limit analysis: comparison of benchmark slope stability problems. ASCE J. of Geotechnical and Geoenvironmental Engineering, DOI: 10.1061/(ASCE) GT.1943-5606.0001347.

[26] Shukla, S.K., Sivakugan, N. and Das, B.M. (2011). A State-of-art review of geosynthetic-reinforced slope. Int. J. of Geotechnical Engineering, 5, p. 17-32

[27] Gao, Y., Yang, S., Zhang, F. and Leshchinsky, B (2016). Three dimensional reinforced slopes: evaluation of required reinforcement strength and embedment length using limit analysis. Geotextiles and Geomembranes,44, p. 133-142.

[28] Song F. Chen RY, Ma LQ & Cao G. (2016). A new method for the stability analysis of geosynthetic-reinforced slopes. Journal Mountain Science 13(11). DOI: 10.1007/s11629-016-4001-8

[29] Christopher, B.R., Zornberg, JG, Mitchell, J.K. (1998). Design Guidance for reinforced soil structures with Marginal soil backfills. Proc. of Sixth Int. Conf. on Geosynthetics, Atlanta, Georgia, March, Vol. 2, 797-804.

[30] Naughton, P.J., Jewell, R.A. & Kempten, G.T. (2001). The design of steep slopes constructed from cohesive fills and a geogrid. Proc. of the International Symposium on Soil Reinforcement, IS Kyushu, Japan, p. 259-264.

[31] Clancy, J.M. & Naughton, P.J. (2008). Design of Reinforced soil structures using fine grained fill types. Advances in Transportation Geotechnics - Ellis, Yu, McDowell, Dawson & Tom (eds.), Taylor & Francis, London, ISBN 978-0-415-475907

[32] Giroud, J.P., Naughton, P.J., Rimoldi, P. & Scotto, M. (2014). Design of reinforced slopes and walls with low-permeability fills using draining geogrids. Proc. of10 Int. Conf. on Geosynthetics, Paper No. 248, September, Berlin Germany, 9 pages.

[33] Abd, A.H. and Utili, S. (2017). Design of geosynthetic reinforced slope

in cohesive backfills. Geotextiles and Geomembranes, 45, p. 627-641.

[34] Ingold, T.S.C. (1982). An Analytical study of Geotextile Reinforced Embankment. Proc. 2nd Int. conf. on Geotextiles, Las Vegas, IFAI, p. 683-688

[35] Leshchinsky, D. and Reinschmidt, A.J. (1985). Stability of membrane reinforced slopes. J. Geotech. Engrg, 111(11), p. 1285-1300.

[36] Wright, S.G. and Duncan, J.M. (1991). Limit equilibrium stability analyses for reinforced slopes. Transportation Research Report, No. 1330, p. 40-46.

[37] Mandal, J.N. and Labhane, L. (1992). A procedure for the design and analysis of geosynthetic reinforced soil slopes. Geotech. and Geological Engrg., 10(4), 291-319.

[38] Srbulov, M. (2001). Analysis of stability of geogrid reinforced steep slopes and retaining walls. Computers and Geotechnics, 28(4), p. 255-268.

[39] Shahgholi, M. & Fakher, A. and Jones, C.J.F.P. (2001). Horizontal slice method of analysis. Geotechnique 51(10), p. 881-885.

[40] Sawicki, A. and Lesniewska, D. (1989). Limit analysis of cohesive slopes reinforced with geotextiles. Computers and Geotechnics,7, p. 53-66.

[41] Michalowski, R.L. and Zhao, A. (1993). Failure criteria for homogenized reinforced soils and application in limit analysis of slopes. Proceedings of Geosynthetics '93. Vancouver, Canada, p. 443-453.

[42] Michalowski, R.L. and Zhao, A. (1994). The effect of reinforcement length and distribution on safety of slopes. Proceedings of the 5th International Conference on Geotextiles, Geomembranes and Related Products. Singapore, p. 495-498.

[43] Michalowski, R. L. and Zhao, A. (1995) Continuum versus structural approach to stability of reinforced soil structure. J. of Geotechnical Engineering Division, ASCE, 121, p. 152-162.

[44] Jiang, G.L. and Magnan, J.P. (1997). Stability analysis of embankments: comparison of limit analysis with method of slices. Geotechnique, 47(4), p. 857-872.

[45] Rowe, K. and Soderman, K. L. (1985). An approximate method for estimating the stability of geotextile-reinforced embankments. Can. Geotechnical J., 22(3), p. 392-398.

[46] Almeida, M.S.S., Britto, A.M. and Parry, R.H.G. (1986). Numerical modeling of a centrifuged embankment on soft clay. Can. Geotechnical J., 23, p. 103-114.

[47] Chalaturnyk, R.J., Scott, J.D., Chan, D.H.K. and Richards, E.A. (1990). Stresses and Deformations in a Reinforced Soil Slope. Can. Geotechnical J., 28, p. 393-402.

[48] Ali, F.H. and Tee, H.E. (1990). Reinforced Slopes: field behavior and prediction. Proc. of the 4th Int. Conf. on Geotextiles, Geomembranes and Related Products, Hague, Netherlands, p. 17-20.

[49] Griffiths, D.V. and Lane, P.A. (1999). Slope stability analysis by finite element method. Geotechnique, 49, 3, p. 387-403.

[50] Krahn, J. (2003). The 2001 R.M. Hardy Lecture: The limits of limit equilibrium analyses. Can. Geotech. J, 40, p. 643-660.

[51] Koerner, R. M. (1990). Designing with Geosynthetics, 2nd Ed., Englewood Cliffs, N.J. Prentice Hall

[52] Fredlund, D.G., and Krahn, J. (1977). Comparison of slope stability methods of analysis. Can. Geotechnical J., 14, p. 429-439.

[53] Fredlund, D.G. (1984). Analytical methods for slope stability analysis. Proc. of 4th Inter. Symposium on landslides, State-of-the-Art, Sep. 16-21, Toronto, Canada, p. 229-250.

[54] Reddy GVN, Madhav MR, Reddy ES (2008). Pseudo-static seismic analysis of reinforced soil wall effect of oblique displacement. Geotextile & Geomembrane, 26(5): p. 393-403.

[55] Zornberg, J.G. and Mitchell, J.K. (1994). Reinforced soil structures with poorly draining backfills, Part I: Reinforcement Interactions and Functions. Geosynth. Int., 2, p. 103-147.

[56] Mitchell, J.K and Zornberg, J.G. (1995). Reinforced soil structures with poorly draining backfills Part II: Case Histories and Applications. Geosynthetics International, 2(1), p. 265-307.

[57] Naughton, P.J., Giroud, J.P., Rimoldi, P., Scotto, M. and Crowther, D. (2015). The design of steep slopes using low-permeability fill and draining geogrids. Proc. XVI ECSMGE, ISBN 978-0-7277-6067-8

[58] Giroud, J.P., 1983. Geotextile drainage layers for soil consolidation. Civil Engineering for Practicing and Design Engineers, Vol. II, p. 275-295.

[59] Michalowski, R.L., 2013. Stability assessment of slopes with cracks using limit analysis. Can. Geotechnical J. 50 (10), p. 1011-1021.

[60] Bishop, A. W. & Morgenstern, N. R. (1960). Stability coefficients for earth slopes. Geotechnique, 10, p. 129-150.

[61] Jha A.K., Madhav Madhira and Reddy G.V.N. (2018). Analysis of Effect of Reinforcement on Stability of Slopes and Reinforcement Length optimization. Geotechnical Engineering Journal of the SEAGS & AGSSEA, Vol. 49 No. 4 December 2018 ISSN 0046-5828

[62] Jha A.K., Madhira, M. and Reddy G.V.N. (2018). Mechanics of Reinforcement – Slope Interactions. INAE Letters, An official Journal of the Indian Academy of Engineering, ISSN 2366-326X, DOI 10.1007/s41403-018-0040-5

[63] Jha A.K., Madhira, M. and Reddy G.V.N. (2016). Analysis of Effect of Reinforcement on stability of Slopes. Indian Journal of Geosynthetics and Ground Improvement, 2018, ISSN 2777-5633, 7, 1, p. 22-27.

[64] Jha A.K., Madhira, M. and Reddy G.V.N. (2017). Analysis of Effect of Curtailment of Reinforcement on Stability of Steep Slopes. Proc, 19th International Conference on Soil Mechanics and Geotechnical Engineering, Seoul 2017, p. 2159-2162

[65] Pockoski, M. and Duncan, J.M. (2000), "Comparison of Computer Programs for analysis of Reinforced slopes", Report of a study performed by the Virginia Tech Center for Geotechnical Practice and Research, Blacksburg, VA 24061.

[66] Liu, C. N., K. H. Yang, Y. H. Ho, and Chang, C. M. (2012). Lessons learned from three failures on a high steep geogrid-reinforced slope. Geotextiles and Geomembranes, 34, p. 131-143.

[67] Yang, K.-H., Thuo, J. N., Chen, J.-W. and Liu, C.-N. (2018). Failure investigation of a geosynthetic-reinforced soil slope subjected to rainfall. Geosynthetics International. [https://doi.org/ 10.1680/jgein.18.00035].

Chapter 3

Design Techniques in Rock and Soil Engineering

Zahid Ur Rehman, Sajjad Hussain, Noor Mohammad, Akhtar Gul and Bushra Nawaz

Abstract

At the initial stage of tunnel design, the tunnel stability can be assessed by different design techniques which are broadly classified into three categories i.e. Mathematical Analysis, Empirical Methods and Numerical Analysis. Mathematical methods or closed form solutions are more precise methods; however, its use is limited to simple geometries and almost impossible for complex geometries due to complex and tedious calculations involved. In practice, Empirical and Numerical Methods are usually used for stability analysis of tunnels. It should be noted that it is not the replacement of final design. Empirical design methods use information about the structural geology and other rock mass properties as input that can be easily obtained at the initial stage of a project. Numerical Methods commonly require mechanical properties, especially strength and deformation of rocks. Numerical methods are also considered as precise due to provision of allowance for variable inputs and geometry and having ability for sensitivity analysis. It is good practice to evaluate the stability of tunnels using at least two Empirical methods and validated through Numerical methods.

Keywords: tunnel design, design techniques, stability, sensitivity, RMR

1. Introduction

The process of engineering design comprises of devising a scheme/module, or process to achieve the required goal or target. It can also be defined as an assessment –making procedure, which utilized the knowledge of basic sciences, mathematics and engineering sciences to convert resources optimally to meet quantified objectives. In other words, engineering design is the procedure of formulating framework, segment, or procedure to address desired problems [1]. General goal of engineering design is to develop a solution (the design) to a known problem. However, there is no single solution, and depends upon the approach used by different engineers resulting different solution. Among the solution obtained some will work well than others, but it is necessary that all solutions should 'work'. The reason behind the fact that solutions to engineering design problem are not unique is perhaps due to very broader spectrum of the concerns encountered in design [2].

2. The design process

Each and every engineering problem/task passes through a design process. According to Hill (1983), as discussed by Biniawski (1988), the design process is:

a) logical development of design inside organization of actions and b) a work plan process for planning the design program. For satisfactory design results, a define process can work as agenda of activities. The defined process or methodology can be considered as a form of quality control that ensures that all aspects that should be considered in the design are considered [2]. Response to a complex engineering problem does not shortly seem in a vacuum. Well-meaning description of engineering problem needs exercise or approach. Design processes generally depend upon the number of engineers analyzing design. The process described here is general, and one can adapt it to the problem, they are trying to solve [3]. Following are the different stages of design process [1] illustrated in **Figure 1**.

1. Recognition of need or a problem
2. Statement of the problem
3. Collection of information
4. Analysis of solution component
5. Synthesis to create a detailed solution
6. Evaluation of ideas and solutions
7. Optimization
8. Recommendation
9. Communication
10. Implementation

2.1 Recognition of need or a problem

Engineering design activity always occurs in response to a human need [3]. Before attempting any solution for design, the presence and nature of a problem must be. This is not an easy task. It needs the rather rare skill of inquiring the right kind of question and call for a clear identification of the problem to be solved. In design it involves the recognition of a genuine social need want or opportunity.

2.2 Statement of the problem

If there is any problem involves, it is then necessary to clearly define it. This may involve a list of specification or criteria. These must be stated clearly and concisely. A poorly recognized and expressed problem cannot be anticipated to result in a good solution. In rock mechanics design, this means to set design objectives in terms of economy, safety and stability.

2.3 Collection of information

This stage comprises the collecting, investigation, processing and analyzing of information to obtain the explicit nature of the targeted problem. In rock engineering collection of information include site investigations, conducting in-situ and

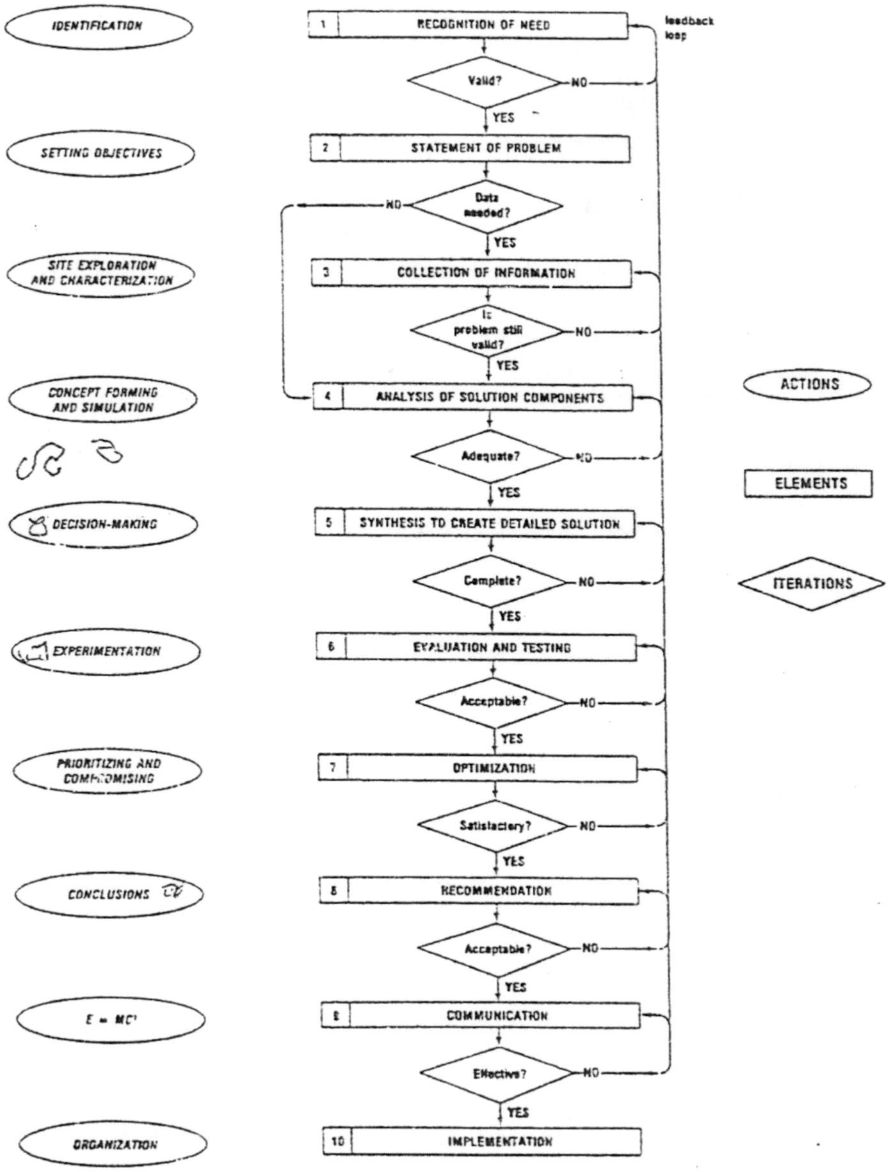

Figure 1.
The engineering design process [1].

laboratory tests to determine the characteristics of the rock strata and assessment of applied loads and field stresses.

2.4 Analysis of solution component

The selection of approach to either search for the most promising method of solution or certain hypothesis is selected or conceived depends upon the nature of the problem. Design approaches at this phase involve numerical analysis and mathematical, physical modal studies, observation and monitoring or the empirical analyses based on experience.

2.5 Synthesis to create a detailed solution

On the basis of analysis of the individual solution component, all design is focused to furnish comprehensive alternative solutions. In this phase of design, calculations, specifications, performance predictions, cost estimates, scheduling procedures and the experimentation are involved.

2.6 Evaluation of ideas and solutions

In this phase the solution is interpreted and compare with the original hypothesis, specification, facts assumptions, requirements or constraints. This demand for a clear understanding of the all relevant interacting factors that's needed for the engineering judgments. The solution for engineering problems should be balanced involving all the factors with interact.

2.7 Optimization

Optimization is the assortment of a best solution (with regard to some criteria) from some set of available alternative solutions [4]. There are always multiple solutions available to any engineering problem. Refinement and modification of a solution may then be required to reach a practicable agreement between the generally contradictory constraint and assets. The effectiveness of an optimization process mostly depends upon simplicity and clarity with which problem and solution are specified.

2.8 Recommendation

Recommendation is the principle of the whole Engineering design process. It provides a refined endorsement of the solution to problem, point out limitations and shows the trend to be followed in applying the solution.

2.9 Communication

The conclusive aim of the all design stages is the creation or instigation of a progression accomplishment. In order to achieve the objective requires the engineer must communicate the finding effectively. Effective communication means that all relevant aspects should be appropriately presented. If a mathematician were to sum up these thoughts, he might well do so by the Eq. (1).

$$E = MC^2 \tag{1}$$

Where,

E means effectiveness of the subject, M mean the mastery of the subject matter and C means the communication.

So for effective communication one should have sound knowledge of the subject matter and good communication skills. The design engineer must have the capability to communicate views and ideas concisely and clearly and to convey technical knowledge effectively.

2.10 Implementation

This is the final stage of design procedures. The finding or results communicated are applied under the given circumstance and proper monitoring is carried out for further refining the result or design that has been recommended for action. The main objective of the design is to ensure that a desire goal and quality will achieved within the time frame and the budget allocated.

2.11 Feed back

After implementation of the design, its performance is monitored and recorded. Remedial measurements are suggested for more improvement of the performance the solution design.

3. Design techniques in soil and rock engineering

There are different significant design techniques in rock engineering. They are classified into three groups which are Analytical, Empirical and Observational. Rock masses having more complex in nature. Due to the very complex nature of rock masses and the difficulties encountered with their characterization, the analytical approach is the least used in the present engineering practice. Due to this reason, it does not lie in the analytical techniques themselves, since some have been developed to a high degree of sophistication, but in the inability to furnish the necessary input data as the ground conditions are adequately explored. Consequently, such analytical techniques as the finite element method, the boundary element method, closed form mathematical solutions, photo-elasticity or analogue simulation are mainly useful for assessing the influence of the various parameters or processes and for comparing alternative design schemes; they are the methods of the future not as yet acceptable as the practical engineering means for the design of rock tunnels [5]. Empirical methods of design are commonly applied as these are built on earlier practices derived from creation of rock structures owning alike physical characteristics [6]. It is a good practice to evaluate the stability of tunnels using at least two Empirical methods and validate through Numerical methods. Therefore, these two groups of tunnel design methods will be discussed in detail [7].

4. Empirical methods of design

The empirical approach relates the experience encountered at previous projects to the conditions anticipated at a proposed site. If an empirical design is backed by a systematic approach to ground classification, it can effectively utilize the valuable practical experience gained at many projects, which is so helpful to exercising one's engineering judgment. This is particularly important since, a good engineering design is a balanced design in which all the factors which interact, even those which cannot be quantified, are taken into account; the responsibility of the design engineers is not to compute accurately but to judge soundly. Rock mass classifications, which the main part of the empirical design methods, are extensively used tunnels within rock. At present, most of the tunnels excavated in the United States make use of some classification system. Terzaghi classification which was presented over 40 years ago is the most broadly used. In fact, rock mass classifications have been successfully applied throughout the world [5].

The empirical methods of design may be used in association with other engineering assessment and design Techniques [6]. These methods are very essential and beneficial for the design in the earlier stages of the project, when minimum evidence about the behavior of rock mass, stress conditions and hydrological characteristics are obtainable [8].

4.1 Rock mass classification systems

Rock mass classification is a tool for the assessment of the rock behavior and performance based on the essential inherent and structural parameters [9]. Rock mass classification systems are the most and widely used empirical methods of design Different rock mass classification systems are RMR, Q-System, RQD, RSR, GSI etc. [6]. Rocks have been classified on the basis of origin, mineralogical compositions and distinct physical properties and ground condition. Rock Classification provides a mutual basis of communication to recognize rock mass in a category having same and well define characterization and basic input parameters for rock engineering design. For designing purposes in several attempts were made to classify rock based on rock and site characterization. Such simplified classification systems have served to understand the upper bound response of the rocks [10]. Rock mass classification systems effectively combined the results comes observation, experience and other engineering judgment for providing a quantitative evaluation of rock mass situations. Rock mass classification systems has the below mentioned purposes in tunneling design [5].

1. Group rock masses having similar behaviors.

2. Provides the root for understanding the characteristics of independent groups.

3. Helps in planning and designing of excavation in rock and provide quantifiable data for the design of complex engineering complications.

4. A common understanding agenda for all the related people in the project.

Up till now different rock mass classification systems have been proposed by Terzaghi (1946), Lauffer (1958), Deere (1964), Wickham, Tiedemann, and Skinner (1972), Bieniawski (1973), and Barton, Lien, and Lunde (1974), (Bieniawski Z. T. 1990). The different classification systems used for the design purposes are assembled in **Table 1**.

4.1.1 Terzaghi's rock mass classification

A well-known classification system for support of tunnels. This explanatory system was developed in the U.S.A in 1946. Terzaghi's (1946) formulate the first rational method of evaluating the rock loads suitable to the design of steel sets. This classification is appropriate for the estimating rock loads for steel arch supported tunnels. It is not so suitable for modern tunneling methods using shotcrete and rock bolts [5].

Terzaghi's classify rocks as under [11]:

1. ***Intact Rock***: Rocks that's having no joints and cracks, it breaks crossways a sound rock or loose block may drops off the top for many hours and days due to blasting. It is called sapling condition.

   Stratified rock: that rock composed those distinct sections having slightly or no confrontation to parting beside the margins stuck between the strata. In such rock the spalling condition is generally happened.

2. ***Moderately jointed rock:*** That rock having joints and hair cracks, but the blocks among joints are locally developed collectively or so closely joined that perpendicular walls do not need on the sides support. In this type of rock, both spalling and popping conditions may be happened.

S. No	Rock mass classification system	Originator	Origin country	Application areas
1	Rock Load	Terzaghi, 1946	USA	Tunnels with steel support
2	Stand-up time	Lauffer, 1958	Australia	Tunneling
3	New Austrian Tunneling Method (NATM)	Pacher et al., 1964	Austria	Tunneling
4	Rock Quality Designation (RQD)	Deer et al., 1967	USA	Core logging, Tunneling
5	Rock Structure Rating (RSR)	Wickham et al., 1972	USA	Tunneling
6	Rock Mass Rating (RMR) Modified Rock Mass Rating (M-RMR)	Bieniawski 1973 (List modified, 1989-USA) Özkan and Ünal, 1990	South Africa Turkey	Tunnels, Mines, (Slopes, Foundations) Mining
	Rock Mass Quality (Q)	Barton et al., 1974 (Last modified 2002)	Norway	Tunnels, Mines, Foundations
8	Strength- Block Size	Franklin, 1975	Canada	Tunneling
9	Rock Mass Strength (RMS)	Stille et al., 1982	Sweden	Metal Mining
10	Unified Rock Mass Classification System (URMC)	Williamson, 1984	USA	General Communication
11	Weakening Coefficient System (WCS)	Singh, 1986	India	Coal Mining
12	Basic Geotechnical Classification	ISRM, 1981	International	General
13	Geological strength index (GSI)	Hoek et al. 1995		Mines and Tunnels

Table 1.
Most widely used rock mass classification systems [6, 10].

3. Blocky and seamy rock: Such rocks consist of chemically intact or almost intact rock fragments which are totally detached from each other and erroneously joined. In such rock, vertical walls may need sides support.

4. Crushed rock: such rocks are chemically intact rock but have the characteristic of crusher outing. If maximum or completely all the fragments are as small as fine sand particles and no cementation has taken place, crushed rock below the water table demonstrate the properties of water-bearing sand.

5. Squeezing rock: Squeezing rock gradually progresses into the tunnel without noticeable increase in volume. An obligation for squeeze is a high percentage of microscopic and sub-microscopic elements of micaceous minerals or clay minerals with a low swelling capability.

6. Swelling rock: Such rock moves inside the tunnel mainly because of expansion. The capability to swell seems to be insufficient to those rocks that have clay minerals such as montmorillonite, with a high swelling capability.

4.1.2 Classifications containing stand-up time

Lauffer (1958) anticipated that stand up time for an excavation span is associated with the quality of rock mass in which the width is mined. The Unsupported span may be defined as the width of the tunnel or the distance between the face and the adjacent support, if such is grater that the tunnels width. Laufer's (1958) advanced classification has been improved by various researchers especially Pacher et al., (1974) and currently formulae the part of the worldwide tunneling attitude so called the New Austrian Tunneling Method (NTAM). The importance of the standup time is to increase in the tunnel width results in a substantial decrease in the period available for the fixing of support. The NATM comprises numerous systems for workable, safe and stable excavation in rock situations where the stand-up time is restricted before collapse occurred. These systems are:

- The use of small headings and benching

- The use of several small drifts to form a reinforced ring inside which the unpackaged of the tunnel can be mined

As described by Terzaghi (1946), these practices are appropriate to apply in squeezing soft rock mass i.e. shale's, phyllites and mudstones. The practices are also appropriate when tunneling in exceptionally jointed rock, but needs excessive attention to apply these practices to underground excavations designed in hard rocks having dissimilar failure mechanisms. For hard rock excavation support design, it is practical to accept the assumption that the stability of the rock mass adjacent to the underground excavation is not time-dependent. A defined wedge visible in the roof of an excavation will fall as soon as after excavation. This can happen after blasting or during the succeeding scaling process. Early support is demanded do keep such a wedge in place, or to improve the limit of safety preferably before the rock supporting the full wedge is removed. On the other hand, in a highly stressed rock condition, failure will generally be induced by some change in the stress condition adjoining the excavation. The failure may occur gradually and apparent it as spalling or it may occur rapidly in the form of a rock burst. In either case, the support system design must take into account the modification in the stress condition rather than the 'stand-up' time of the excavation.

4.1.3 Rock quality designation index (RQD)

It is developed by Deere et al., (1967). Such system provides the quantities estimation of rock mass quality from the drill core logs. RQD is defined as the percentage sum of all intact core pieces having length more than 10 cm in the total length of the core provided that the core should be of NX size (54 mm in diameter). The precise practices for the estimation of the size of core portions and the approximation of Rock Quality Designation Index are summarized as shown in **Figure 2** [11].

In 1982, Plastron suggested that when core is not available and discontinuity traces are visible in surface disclosure or exploratory adits, the RQD may be calculated from the number of discontinuities per unit volume. The suggested relationship is for clay free masses and is given below by Eq. (2).

$$RQD = 115 - 3.33\,J_v \tag{2}$$

Where,

RQD is the Rock Quality Designation Index,

J_v is the number of all joints per unit length for all joint (discontinuity) sets, so called volumetric joint count.

4.1.4 Rock structure rating

Wickham et al. (1972) established another quantitatively rock mass classification system termed as Rock Structure Rating (RSR). RSR is used to describe and measure

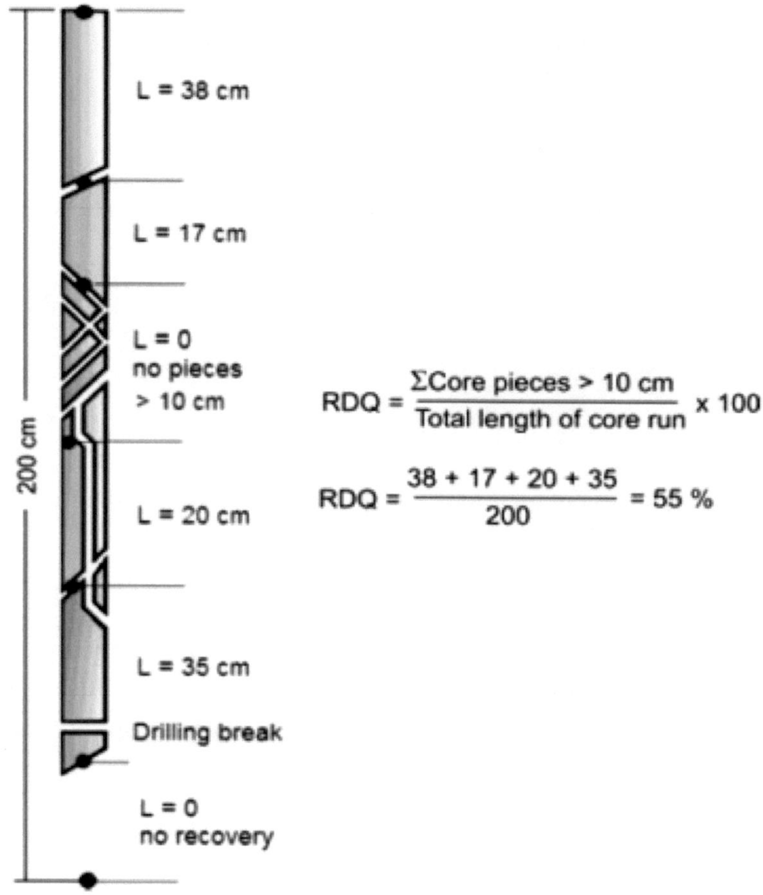

Figure 2.
Procedure for measurement and calculation of RQD [11].

the quality of rock mass for selecting of appropriate support and reinforced system. Such classification system not applied generally as compared to other classification systems, but it has its important role in the emergent of other empirical classification schemes. Many investigators advised that for good, reliable and suitable results for planning of excavation more than one rock mass classification systems should be used at initial stage of the project. The significance of the rock structure rating, in the context of this conversation, is to bring forward the idea of assessment of each of the constituents recorded below to calculate a mathematical value of RSR = A + B + C.

Where,

Factor A: Area Geology: It includes Common evaluation of geological structure based on:

- Rock type Origin (sedimentary, metamorphic and igneous).
- Rock Hardness (it means hard, medium, soft and decomposed).
- Geologic structure (immense, marginally faulted/folded, reasonably faulted/folded, extremely faulted/folded).

Factor B: Geometry of the geological structures: it consists of effect of disjointedness arrangement with consideration to the tunnel alignment on the basis of:

- Joint spaces.
- Orientation of joints (dip and strike).
- Direction of tunnel drive.

Factor C: it includes influence of groundwater inrush and joint situation on the basis of:

- Whole rock mass class based previous parameter combined (A and B).
- Situation of Joint (poor, fair and good).
- Quantity of water flow (gallons/minute/1000 feet of tunnel).

The following tables are used for the calculation of RSR (maximum RSR is 100) [9] (**Tables 2–4**).

	Basic Rock Type				Geological Structure			
	Hard	**Medium**	**Soft**	**Decomposed**				
Igneous	1	2	3	4		Slightly	Moderately	Intensively
Metamorphic	1	2	3	4		Folded or	Folded or	Folded or
Sedimentary	2	3	4	4	Massive	Faulted	Faulted	Faulted
Type 1					30	22	15	9
Type 2					27	20	13	8
Type 3					24	18	12	7
Type 4					19	15	10	6

Table 2.
Rock structure rating, parameter a: General area geology [9].

	Strike ⊥ to Axis					Strike ‖ to Axis		
	Direction of Drive					Direction of Drive		
	Both	With Dip		Against Dip		Either direction		
	Dip of Prominent Joints[a]					Dip of Prominent Joints		
Average joint spacing	Flat	Dipping	Vertical	Dipping	Vertical	Flat	Dipping	Vertical
1. Very closely jointed, < 2 in	9	11	13	10	12	9	9	7
2. Closely jointed, 2–6 in	13	16	19	15	17	14	14	11
3. Moderately jointed, 6–12 in	23	24	28	19	22	23	23	19
4. Moderate to blocky, 1–2 ft	30	32	36	25	28	30	28	24
5. Blocky to massive, 2–4 ft	36	38	40	33	35	36	24	28
6. Massive, > 4 ft	40	43	45	37	40	40	33	34

[a]*Dip: flat: 0–20°; dipping: 20–50°; and vertical: 50–90°.*

Table 3.
Rock structure rating, parameter B: Joint pattern, direction of drive [9].

	Sum of Parameters *A* + *B*					
	13–44			45–75		
Anticipated water inflow gpm/1000 ft. or tunnel	Joint Condition[a]					
	Good	Fair	Poor	Good	Fair	Poor
None	22	18	12	26	22	18
Slight, < 200 gpm	19	15	9	23	19	14
Moderate, 200–1000 gpm	15	22	7	21	16	12
Heavy, > 1000 gp	10	8	6	18	14	10

[a]*Joint condition: good = tight or cemented; fair = slightly weathered or altered; poor = severely weathered, altered or open.*

Table 4.
Rock structure rating, parameter C: Groundwater, joint condition [11].

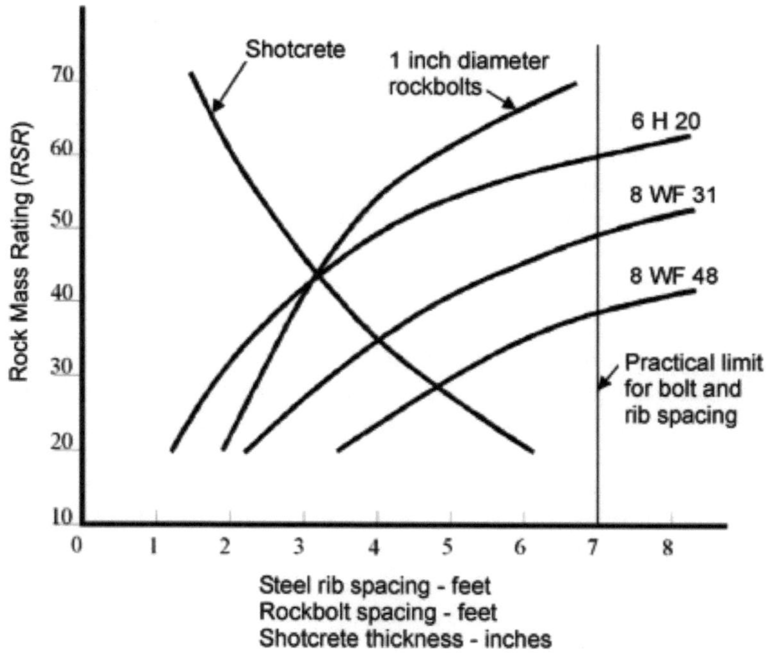

Figure 3.
RSR support recommendation chart [9].

The RSR value calculated for the above tables are then used for the calculation support system recommendation. The support recommendation chart for the RSR value is given in **Figure 3**.

4.1.5 Rock mass rating system (RMR system)

The rock mass rating system was produced by Biniawski in 1976; it is sometimes also called geo-mechanics classification system. It was developed taking into account the distinctive case histories in the field of structural designing This classification system was altered in 1974, 1976, 1979 and 1989, because of considering of more contextual analyses identified related to tunnels, mines, chambers, slopes and foundations [1]. The Geo-mechanics classification system has a widespread application in different rock engineering fields such as mining, hydro power projects, tunneling and hill slope stability (Kumar S. S., 2012). The geo-mechanics classification incorporates the following 6 parameters that are computable in the site and from cores [6]:

1. Uniaxial compressive strength

2. Rock quality designation (RQD)

3. Spacing of discontinuities

4. Condition of discontinuities

5. Ground water condition

6. Orientation of discontinuities

While using this classification system, the rock masses are divided into a number of structural regions. Each region is classified independently [12]. These six parameters are being given different rating based on different geological and geotechnical condition as shown in **Table 5**.

Based on the overall rating of RMR calculated form above mentioned parameters support systems are being recommended for the project site. Support recommendation based on RMR value is given in **Table 6**.

4.1.6 Q-system

This system of rock mass classification was devised by Barton et al., (1979) in Norwegian Geotechnical Institute (NGI), explicitly for the design of tunnel established on 212 case histories. The rock mass classification system is generally used for tunnel design throughout the world and has been used in approximately 1260 various projects and considered as one of the best classification systems for design of tunnels (Kumar N., 2002). The extreme ratings of Q-System shows good quality of rock mass and the lowest ratings designate poor quality of rock mass. The minimum and maximum of Q-index ranges from 0.001 to 10000 on logarithmic scale. According to this classification system Q is the function of six independent parameters as defined by Eq. (3).

$$\mathrm{Q} = \frac{RQD}{Jn} \times \frac{Jr}{Ja} \times \frac{Jw}{SRF} \tag{3}$$

Where,

RQD Rock Quality designation index, *Jn* shows joint set number, *Jr* shows number of joint roughness estimated for the set of joint that is most terrible and dangerous to alignment of tunnel, *Ja* show joint alteration number estimated for the most dangerous and unfavorable set of joint along the alignment of tunnel, *Jw* is joint water condition which shows the water reduction factor, Stress Reduction Factor, SRF is comprised to consider the consequence of in-situ stress condition on the whole quality of Rock. The following comments are offered by Barton et al. (1974) for explaining the meaning of the parameters used to decide the value of Q.

The first quotient $\left(\frac{RQD}{Jn}\right)$ demonstrating the organization of the rock mass, is a rough measure of the block size.

A. CLASSIFICATION PARAMETERS AND THEIR RATINGS*

	Parameter		Range of values						
1	Strength of intact rock material	Point-load strength index	>10 MPa	4–10 MPa	2–4 MPa	1–2 MPa	For this low range - unlaxial compressive test is preferred		
		Unlaxial comp. Strength	>250 MPa	100–250 MPa	50–100 MPa	25–50 MPa	5–25 MPa	1–5 MPa	< 1 MPa
		Rating	15	12	7	4	2	1	0
2		Drill core Quality *RQD*	90% - 100%	75% - 90%	50% - 75%	25% - 50%	<25%		
		Rating	20	17	13	8	3		
3		Spacing of	> 2 m	0.6–2. m	200–600 mm	60–200 mm	< 60 mm		
		Rating	20	15	10	8	5		
4		Condition of discontinuities (see E)	Very rough surfaces Not continuous No separation Unweathered wall rock	Slightly rough surfaces Separation <1 mm Slightly weathered walls	Slightly rough surfaces Separation <1 mm Highly weathered walls	Slickensided surfaces or Gouge <5 mm thick or Separation 1–5 mm Continuous	Soft gouge >5 mm thick or Separation >5 mm Continuous		
		Rating	30	25	20	10	0		
5	Groundwater	Inflow per 10 m tunnel length (Mm)	None	< 10	10–25	25–125	> 125		
		(Joint water press)/(Major principal σ)	0	<0.1	0.1, – 0.2	0.2–0.5	>0.5		
		General conditions	Completely dry	Damp	Wet	Dripping	Flowing		
		Rating	15	10	7	4	0		

B. RATING ADJUSTMENT FOR DISCONTINUITY ORIENTATIONS (See F)						
Strike and dip orientations		Very favorable	Favorable	Fair	Unfavorable	Very Unfavorable
Ratings	Tunnels & mines	0	−2	−5	−10	−12
	Foundations	0	−2	−7	−15	−25
	Slopes	0	−5	−25	−50	
C. ROCK MASS CLASSES DETERMINED FROM TOTAL RATINGS						
Rating		100 ← 81	80 ← 61	60 ← 41	40 ← 21	<21
Class number		I	II	III	IV	V
Description		Very good rock	Good rock	Fair rock	Poor rock	Very poor rock
D. MEANING OF ROCK CLASSES						
Class number		I	II	III	IV	V
Average stand-up time		20 yrs. for 15 m span	1 year for 10 m span	1 week for 5 m span	10 hrs for 2.5 m span	30 min for 1 m span
Cohesion of rock mass (kPa)		>400	300–400	200–300	100–200	<100
Friction angle of rock mass (deg)		>45	35–45	25–35	15–25	<15
E. GUIDELINES FOR CLASSIFICATION OF DISCONTINUITY conditions						
Discontinuity length (persistence)		<1 m	1–3 m	3–10 m	10–20 m	>20 m
Rating		6	4	2	1	0
Separation (aperture)		None	<0.1 mm	0.1–1.0 mm	1–5 mm	>5 mm
Rating		6	5	4	1	0
Roughness		Very rough	Rough	Slightly rough	Smooth	Slickensided
Rating		6	5	3	1	0

Infilling (gouge)	None	Hard filling < 5 mm	Hard filling >5 mm	Soft filling <5 mm	Soft filling >5 mm
Rating	6	4	2	2	0
Weathering	Unweathered	Slightly weathered	Moderately weathered	Highly weathered	Decomposed
Ratings	6	5	3	1	0
F. EFFECT OF DISCONTINUITY STRIKE AND DIP ORIENTATION IN TUNNELING[**]					
Strike perpendicular to tunnel axis				Strike parallel to tunnel axis	
Drive with dip - Dip 45–90°	Drive with dip - Dip 20–45°			Dip 45–90°	Dip 20–45°
Very favorable	Favorable			Very unfavorable	Fair
Drive against dip - Dip 45–90°	Drive against dip - Dip 20–45°			Dip 0–20 - Irrespective of strike°	
Fair	Unfavorable			Fair	

[*]*Some conditions are mutually exclusive. For example, if infilling is present, the roughness of the surface will be overshadowed by the influence of the gouge. In such cases use A.4 directly.*
[**]*Modified after Wickham et al. (1972).*

Table 5.
Rock mass rating system [5].

Rock mass class	Excavation	Rock bolts (20 mm diameter, fully grouted)	Shotcrete	Steel sets
I. Very good rock *RMR*: 81–100	Full face, 3 m advance.	Generally no support required except spot boiling.		
II. Good rock *RMR*: 61–80	Full face, 1–1.5 m advance. Complete support 20 m from face.	Locally, bolts in crown 3 m long, spaced 2.5 m wiln occasional wire mesh.	50 mm in crown where required.	None.
III. Fair rock *RMR*: 41–60	Top heading and bench 1.5–3 m advance in top heading. Commence support after each blast. Complete support 10 m from face.	Systematic bolts 4 m long, spaced 1.5–2 m in crown and walls with wire mesh in crown.	50–100 mm in crown and 30 mm in sides.	None.
IV. Poor rock *RMR*: 21–40	Top heading and bench 1.0–1.5 m advance in top heading. Install support concurrently with excavation, 10 m from face.	Systematic bolts 4–5 m long, spaced 1–1.5 m in crown and walls with wire mesh.	100–150 mm in crown and 100 mm in sides.	Light to medium ribs spaced 1.5 m where required.
V. Very poor rock *RMR*: < 20	Multiple drifts 0.5–1.5 m advance in lop heading. Install support concurrently with excavation Shotcrete as soon as possible after blasting.	Systematic bolts 5–6 m long, spaced 1–1.5 m in crown and walls with wire mesh. Bolt invert.	150–200 mm in crown, 150 mm in sides, and 50 mm on face.	Medium to heavy ribs spaced 0.75 m with steel lagging and forepoling if required. Close invert.

Table 6.
Guidelines for excavation and support of 10 m span rock tunnels in accordance with the RMR system [1, 6].

The second quotient $\frac{Jr}{Ja}$ communicates the unevenness and frictional features of the joint walls or infill materials. This measure is taken in favor of uneven, unchanged joints in direct interacted. The strength is reduced significantly in case where rock joints have coating of thin clay mineral and fillings. It defines the inter – block shear strength of rock mass.

The third quotient $\frac{Jw}{SRF}$ incorporates two stress related parameters. SRF is a degree of 1) untying load when the excavation passes through clay bearing rock and shear zones, 2) rock stress when the excavation is within competent rock, and 3) squeezing loads in plastic weak rock masses. It is also as a total stress parameter. The Jw parameter is amount of water pressure, adversely affect the shear strength of joints as it reduces the effective normal stress. In addition, presence of water may create softening and ultimately the possibility of outwash when clay infill the joints. It generally shows the active stress component and that is determined empirically. The comprehensive and detail system of determining the values of the Q-System parameters (Rock quality designation *(RQD)*, Number of joints *(Jn)*, Roughness number for joint *(Jr)*, Joint alteration number *(Ja)*, Joint water reduction factor *(Jw)*, Surface reduction factor *(SRF)* are given in **Tables 7–12**. The extreme value exemplifies good class of rock and the inferior value signifies poor class of rock.

The values achieved for the different parameters using the above cited tables are then used for the determination of the value of the Q- system. Based on the Value of Q-System the Bortan et al. (1974) classify the quality of rock into nine different groups as shown in **Table 13**.

1	Rock quality designation (RQD)		RQD
A	Very poor	>27 joints per m^3	0–25
B	Poor	20–27 joints per m^3	25–50
C	Fair	13–19 joints per m^3	50–75
D	Good	8–12 joints per m^3	75–90
E	Excellent	0–7 joints per m^3	90–100

Note: i. Where RQD is reported, as ≤10 (including zero) the value 10 is used to assess the Q-value.
ii. RQD-intervals of 5 are adequately accurate.

Table 7.
Rock quality designation (RQD) and volumetric jointing [13].

2	Jn values	Jn
A	Massive, no or few joints	0.5–0.1
B	One joint set	2
C	One joint set plus random joints	3
D	Two joint sets	4
E	Two joint sets plus random joints	6
F	Three joint sets	9
G	Three joint sets plus random joints	12
H	Four joint sets, random, heavily jointed, "sugar cube", etc.	15
I	Crushed rock, earth like	20

Note: i. For tunnel intersection, use 3 Jn.
ii. Far portals, use 2 Jn.

Table 8.
Joint set numbers (Jn) values [13].

3	Jr values	Jr
a Rock-wall contact and		
b Rock-wall contact before 10 cm shear movement		
A	Discontinuous joints	4
B	Rough or irregular undulating	3
C	Smooth undulating	2
D	Slickensides, undulating	1.5
E	Rough irregular planar	1.5
F	Smooth planar	1
G	Slickensides planar	0.5
Note: i. description refer to small scale features and intermediate scale features, in that order		
c. No-rock wall contact when sheared		
H	Zones containing clay minerals thick enough to prevent rock wall contact	1
I	Sandy, gravely or crushed zone thick enough to prevent rock wall contact	1

3	Jr values	Jr
Note: ii. 1. Add 1.0 if the mean spacing of the relevant joint set is greater than 3 m. iii. Jr. = 0.5 can be used for planar, slickensides joints having lineation, provided that the lineation are oriented for minimum strength.		

Table 9.
Joint roughness number (Jr) values [13].

4	Ja values	φ_T approx.	Ja
a. Rock-wall contact (no filling, just coatings)			
A	Hard impermeable filling firmly healed hard such as epidolite/quartz		0.75
B	Only surface staining with unaffected joint walls.	25–35°	1
C	A little altered joint-walls with Non-softening mineral coatings; sandy particles/ clay free fractured rock, etc.	25–30°	2
D	Silty/sandy clay coatings. Small clay fraction.	20–25°	3
E	Mineral coatings with clay of low friction, such as Mica/Kaolinite etc.	8–16°	4
b. Rock-wall contact before 10 cm shear with a slim mineral filling			
F	Clay-free fragmented rock, sandy particles	25–30°	4
G	Strongly over-consolidated, non-softening, clay mineral fillings (less than 5 mm Continuous thickness).	16–24°	6
H	Medium or low over-consolidation, softening, clay mineral fillings (less than 5 mm continuous thickness).	12–16°	8
I	Swilling clay fillings, i.e., montmorillonite (less than 5 mm continuous thickness).	6–12°	8–12
c. No rock-wall contact due to thick mineral filling even after shear			
J	Zones or bands of crushed rock. Medium or low over-consolidation.	16–24°	6
K	Zones of clay, disintegrated rock Medium or low over-consolidation.	12–16°	8
L	Zones of clay, disintegrated rock. Joint alteration depends on the percentage of swelling clay-size particles.	6–12°	8–12
M	Thick continuous zones of clay or band of clay. Strongly over consolidated	12–16°	10
N	Thick continuous zones of clay. Joint alteration depends on the percentage of welling clay-size particles.	12–16°	13
O	Thick and continuous clay zones. Joint alteration depends on the percentage of swelling clay-size particles.	6–12°	13–20

Table 10.
Joint alteration (Ja) values [13].

High professionalism is required for estimation of the values of parameter used in this system. The poor professional users may face trouble while approximating the score of the parameters and may approximate the lesser value for Q-System, which is considered the weakness of this classification system [14].

The width and altitude of the underground excavations mainly depend on the class of rock mass and considered as significant elements in design of underground excavations. The facet of width or altitude directly disturbs the stability when amplified or declined. To highlight the safety obligation, Bortan et al. (1974) further

5	Jw values	Jw
A	Dry excavation or minor inflow (humid or a few drips)	1.0
B	Medium inflow, infrequent outwash of joint filling (many drips/“rain”)	0.66
C	Jet inflow or higher pressure in competent rock with unfilled joints	0.5
D	Large inflow or higher pressure, considerable outwash of joint fillings	0.33
E	Exceptionally high inflow continuing without perceptible decay. Causes outwash of material and possibly cave in	0.2–0.1
F	Exceptionally high inflow continuing without perceptible decay. Causes outwash of material and possibly cave in	0.1–0.05

Table 11.
Joint water reduction factor (Jw) values [13].

6	SRF values			SRF
a. Weak zones crossing the underground excavation, which may cause loosening of rock mass				
A	Multiple occurrences of weak zones within a short section containing clay or chemically disturbed very loose surrounding rock at any depth, or long section with incompetent rock.			10
B	Multiple shear zones within a short section in competent day-free rock with weak surrounding rock at any depth.			7.5
C	Single weak zone with or without clay or chemical disintegrated rock with depth less than or equal to 50 m.			5
D	Loose, open joints, heavily jointed at any depth			5
E	Single weak zones with or without clay or chemical disintegrated rock with depth greater than 50 m			2.5
	Note: i. Reduce these values of SRF by 25–50% if the weak zones but do not intersect the underground opening			
b. Competent massive rock with stress problems		σc / σ1	σΘ / σc	SRF
F	Low stress, near surface, open joints	>200	<0.01	2.5
G	Medium stress, favorable stress condition	200–10	0.01–0.3	1
H	High stress, very tight structure. Usually good for stability. Depending on stress orientation it may be unfavorable to stability.	10–5	0.3–0.4	0.5–2 2–5*
I	Moderate spalling land/slabbing after greater than one hour in massive rock	5–3	0.5–0.65	5–50
J	Spalling or rock burst after a few minutes in massive rock	3–2	0.65–1	50–200
K	Heavy rock burst and instant active deformation in massive rock	<2	>1	200–400
	Note: ii. For strongly anisotropic virgin stress field (if measured): when 5 ≤ σ1 / σ3 ≤ 10 reduce σc to 0.8 σc, and σΘ to 0.8 σΘ, when σ1 / σ3 > 10 reduce σc to 0.5 σc, and σΘ to 0.5 σΘ. iii. Few case records available where depth of crown below surface is less than span width Suggest SRF increase from 2.5 to 5 for such cases (see H).			
c. Squeezing rock: plastic deformation in incompetent rock under the influence of high pressure			σΘ / σc	SRF
L	Mild squeezing rock pressure		1–5	5–10
M	Heavy squeezing rock pressure		>5	10–20
d. Swelling rock: chemical swelling activity depending on the presence of water			SRF	

6	SRF values	SRF
N	Mild swelling rock pressure	5–10
O	Heavy swelling rock pressure	10–15

Table 12.
Stress reduction factor (SRF) values [13].

Q-System values range	Group	Classes of rock mass
0.001–0.01	3	Exceptionally Poor
0.01–0.1		Extremely Poor
0.1–1	2	Very Poor
1–4		Poor
4–10		Fair
10–40	1	Good
40–100		Very Good
100–400		Extremely Good
400–1000		Exceptionally Good

Table 13.
Rock mass classification based on Q-system [13].

carry the addition of a fresh parameter to Q-System named as excavation support ratio (ESR). The lower value of ESR symbolizes the necessity of great level firmness and vice versa. The ESR is used for the estimation of support system that can be set up to sustain the stability and also associated to the anticipated use of excavation. Incorporating various conditions, different values of ESR are summarized in **Table 14**. Based on the width and altitude of underground excavation, ESR shows the Equivalent dimension that is achieved by means of the Eq. (4) [13].

$$De = \frac{\text{width or altitude in meter}}{\text{ESR}} \tag{4}$$

The support chart proposed by Bortan et al. (1974) as shown in **Figure 4**, is based on the Q-system ratings and equivalent dimension for the endorsement of permanent support system for underground excavations. This chart provides a

7	Excavation types	ESR values
A	Temporary mine openings	3–5
B	Permanent mine openings, water tunnels for hydro power (excluding high Pressure penstocks), pilot tunnels, drifts and headings for large excavations.	1.6
C	Storage rooms, water treatment plants, minor road and railway tunnels, surge Chambers, access tunnels.	1.3
D	Power stations, major road and railway tunnels, civil defense chambers, Portal intersections.	1.0
E	Underground nuclear power stations, railway stations, sports and public Facilities, factories.	0.8

Table 14.
Excavation support ratio (ESR) [13].

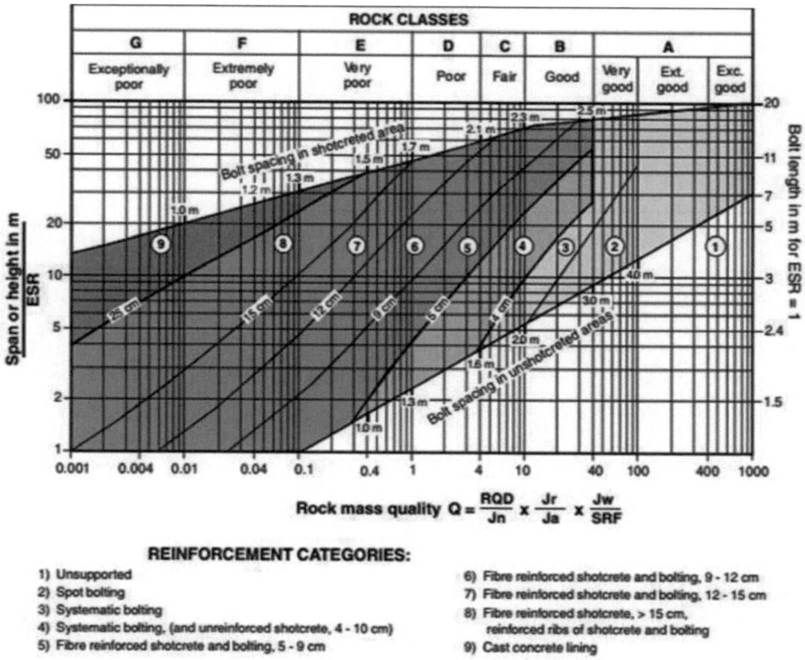

Figure 4.
Permanent support system recommendation chart for Q-system [13].

wide-ranging framework established on the empirical data that what kind of support system is recommended in case of rock bolt's center to center spacing and the thickness sprayed concrete, and also give the energy absorption of fiber strengthened sprayed concrete.

4.2 Geological strength index (GSI)

This classification system established and improved by Hoek and other researchers including the block size and its shear strength in order to estimate value of GSI quantitatively. The GSI index value for any rock mass is depend on the estimation techniques, expertise and reliability of these two input parameters. Sonmez and Ulusay developed the arithmetical basis for GSI value calculation and present quantitatively GSI chart as given in **Figure 5** [16]. Further research were carried out for quantification of GSI value by (Cai, et al.,2004), they present the assessment method for block size, joint and joints wall condition for GSI value quantification.

GSI system should not be considered as the replacement for other classification systems like RMR and Q-System, as this system cannot recommend any support system for stability of rock mass. This system can only be used in estimation of rock mass properties and input parameters for numerical modeling [15]. The comprehensive practice for estimation of input parameters for numerical analysis of stress condition and the remedial measures is presented in **Figure 5** (Hoek, 2013).

The GSI index may be estimated by subsequent various methods used for assessment of rock mass.

Method A: Using this method the GSI is estimated by skilled geologist or mining engineers from the data collected (observational data) at site and then the value of GSI is evaluated from chart [17].

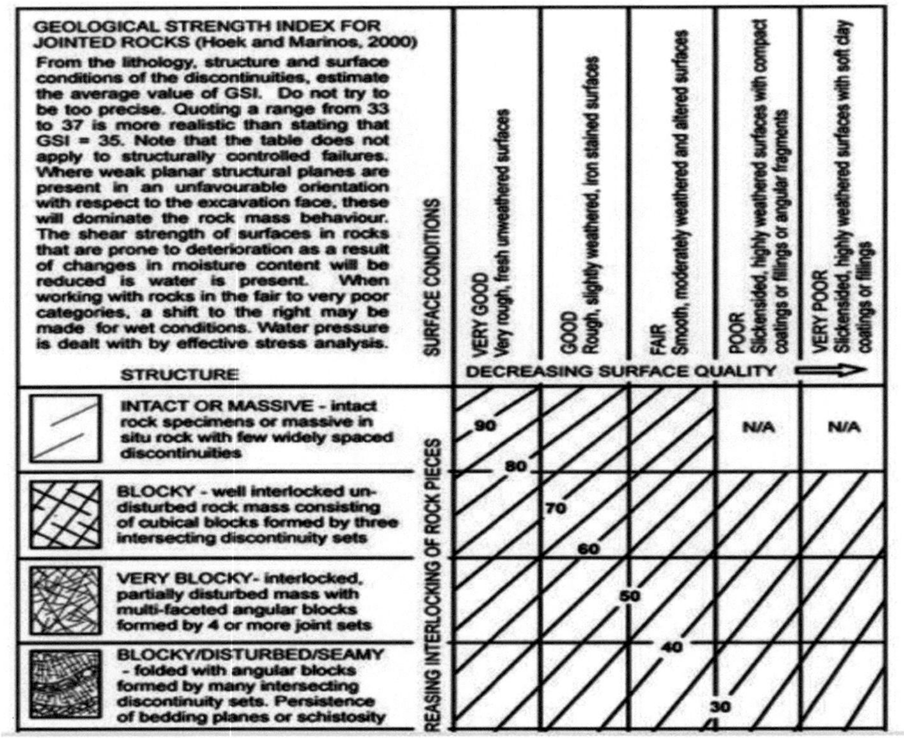

Figure 5.
Geological strength index chart [15].

Method B: In this method the GSI index is estimated by using other classification systems like RQD and RMR etc. when limited data is available. The GSI can be estimate from the well-known relationship presented by various researchers [17].

Method C: The sonmez and Ulusay considered structure rating (SR) and surface condition rating (SCR) for approximation of GSI value [17].

The Cai et al. (2004) used block volume (V_b) and joint surface condition factor (J_c) to approximation the GSI. The block volume having greater number of joint sets indicated as:

$$Vb = \text{S1} \times \text{S2} \times \text{S3} \tag{5}$$

where, S is joint spacing.

The *Jc* defined by the roughness of joint, weathering and infilling, these are used to measure the joint surface condition factor by using the Eq. (6).

$$Jc = \text{Jw} \times \text{Js}/\text{Ja} \tag{6}$$

The V_b and *Jc* are used to precisely quantify the GSI value [17]. The quantitative chart for estimation of GSI suggested by sonmez and Ulusay [1999] is shown in **Figure 6**.

5. Numerical methods of design

The empirical methods of design do not estimate accurately the reliability supports, redistribution of stresses, rock mass deformation [18]. These parameters are very important in designing and analysis of any excavation therefore, numerical

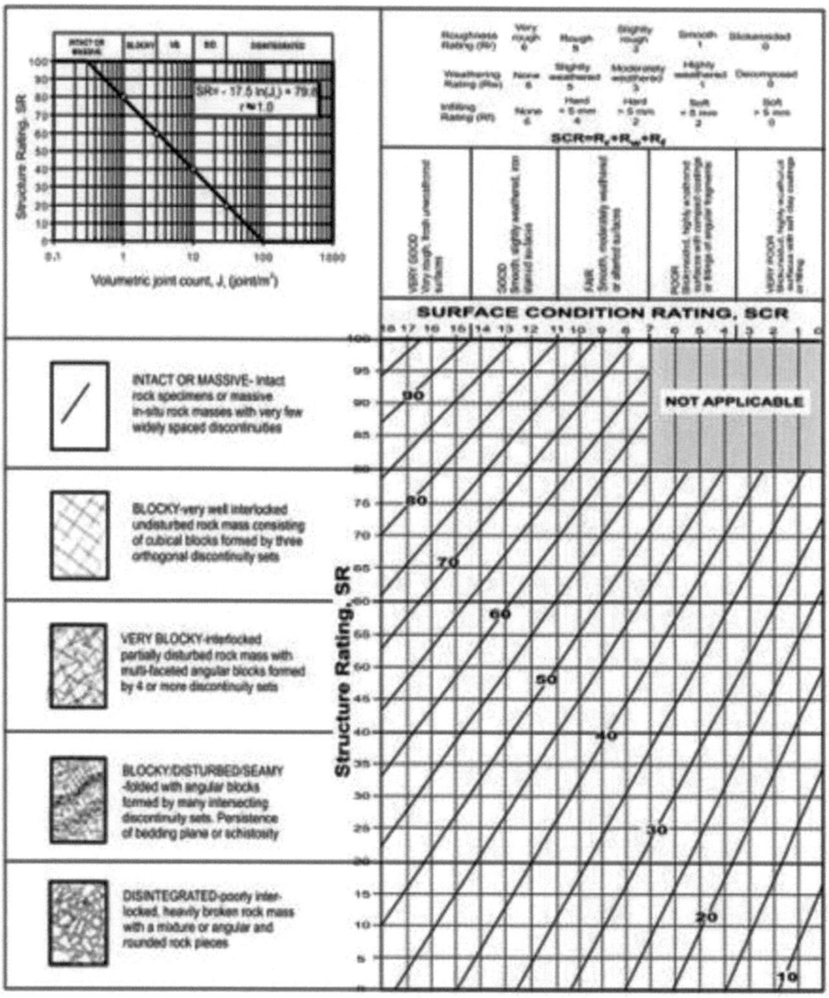

Figure 6.
Quantitative estimation of GSI chart [15].

analysis should be carried out for appropriate designing. The numerical methods are considered very useful to estimate the above parameters precisely and in minimum time as compared to other methods of design. Numerical methods used physical and strength properties of rock as input for analysis. For efficient and viable design the numerical and empirical methods are used in parallel [19–23].

Different researchers developed and present various numerical methods and models. These are divided into eight classes on the basis of four methods and two levels as shown in **Figure** 7 [24, 25].

5.1 Numerical methods of modeling for rock/soil engineering

The numerical methods of design uses in rock/soil engineering are grouped into three classes for modeling in rock mechanics as discussed above.

5.1.1 Continuum methods

The different continuum methods of design are as under.

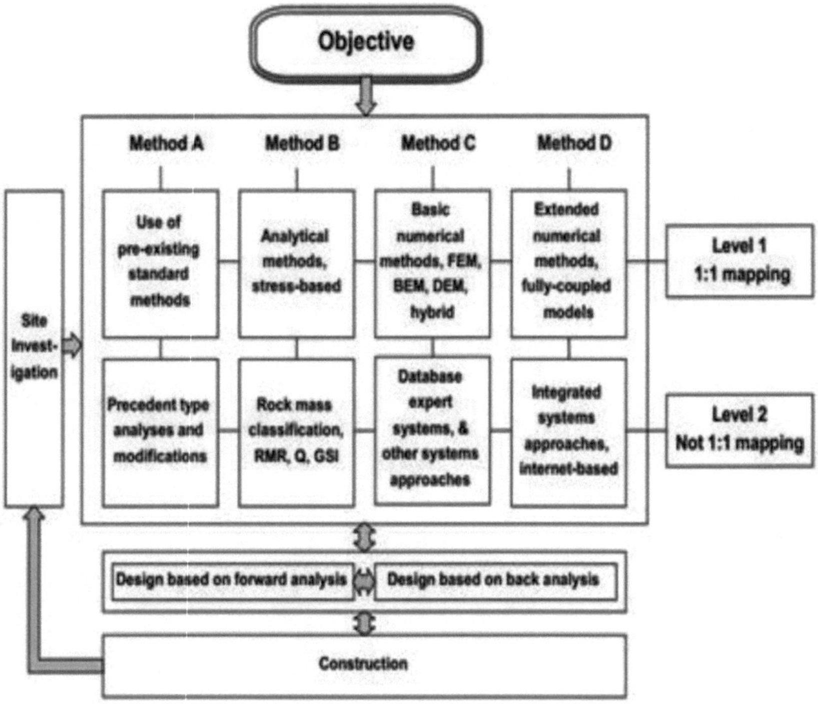

Figure 7.
Division of numerical models and methods [24, 25].

1. Finite Difference Method (FDM)

2. Finite Element Method (FEM)

3. Boundary Element Method (BEM)

Finite Difference Method (FDM).

The Finite difference method (FDM) is the direct calculation of PDEs and transmitted the creative PDEs in term of unknown at grid point into a system of algebraic equations by interchange the fractional derivatives with difference at irregular or regular grid forced over problem areas. This system is solved due to establishing the required initial and boundary condition. This method is old but widely applied in the numerical modeling in rock mechanics. This method is based for explicit approach of discreet element method (DEM) [26].

Finite Element Method (FEM).

The Finite element method (FEM) splits the problem into sub-elements of smaller sizes and shapes with fitting the number of nodes at the vertices and at the side of discretization. FEM is mostly used to estimate the behavior of PDEs at elemental level and for signifying the behavior of elements; it produces the local algebraic equation. After creating the local equation the FEM gathered it according to topographic relation of node and elements and further put it into worldwide system of algebraic equation for receiving the required information after establishing the definite initial and boundary situations.

Boundary Element Method (BEM).

The Boundary element method is the precise method then FEM and FDM because of its easiness. This method involves the discretization of solution areas at boundary and thus decreases the problem dimension by simplifying the design input parameters. This method computes separately the essential information in the solution domains from the information at the boundary, which is achieved by the solution of boundary integral equation rather than direct solution of PDEs [26].

5.1.2 Discontinuum methods

The different discontinuum methods of design are given below.

1. Discrete Element Method (DEM)

2. Discrete Fracture Network (DFN)

5.1.3 Hybrid continuum/Discontinuum

Following are the different Hybrid continuum/discontinuum methods of design:

1. Hybrid FEM/BEM methods

2. Hybrid DEM/DEM methods

3. Hybrid FEM/DEM methods

4. Other hybrid method/models

6. Finite element method (FEM)

This method of design was developed by Clough et al., (1950). Due to wide application of this method in mining engineering especially tunneling, it get more attention for solving mining problems and popularity in this field [19]. The FEM divide problem into small parts and connect these parts at a point/nodes at the apexes and at the boundaries of meshing/discretization. The FEM has many applications in modeling in rock engineering design due to dealing with nonlinearity, boundary conditions and heterogeneity problems [26, 27].

The unidentified function over each element in FEM estimated through test function having its nodal values of anonymous system (in polynomial form). This practice is the fundamental supposition of FEM. For experimental function, it is mandatory to satisfy the principal of PDFs. In this research the FEM based software Phase2 was used for analysis of stresses and total displacement around tunnel. For experimental function it must be satisfied the principal of PDFs, which is given in Eq. (7).

$$u_i^e = \sum_{j=1}^{M} N_{ij} u_i^e \tag{7}$$

Where,

Nij is the shape function or interpolation function; this must be defined into inherent coordinates for use of Gaussian quadratic integration, M is the element order.

Using shape function the problem original PDFs can be substituted by the arithmetical equation as given below.

$$\sum_{j=1}^{N} \left[K_{ij}^e\right] \{u_i^e\} = \sum_{j=1}^{N} f_i^e \quad or \quad Ku = F \tag{8}$$

Where,

Keij is the coefficient matrix, u_i^e vector is the nodal value vector having unidentified variables, f_i^e is consist of body force contribution and initial boundary condition, *K* is the global stiffness matrix.

Keij is also called the element stiffness matrix in term of elasticity problem which is given by Eq. (9).

$$\left[K_{ij}^e = \int_{\Omega i} ([B_i][N_i])^T [D_i][B_i] d\Omega \right. \tag{9}$$

Where,

D_i is the elasticity matrix; B_i is the geometry matrix which is determined from the relation between displacement and strain.

In FEM the material properties of different materials can easily feed into FEM by assigning different properties to different elements distinctly.

6.1 Finite elements

The element may be in numerous forms i.e. one dimensional, two dimensional and three dimensional elements. One dimensional element having cross-sectional area and usually denoted by line sections or segment. Two dimensional element fields consist of triangle and quadrilateral. Three dimensional element field described by tetrahedron and parallelepiped. Some element shapes and node position used in two dimensional element fields [28] (**Figure 8**).

6.2 Shape function

It is the displacement within the element at any point when related to the displacement of the nodes. For instance the displacement of u and v within the quadrilateral element at any point represented by Eq. (10).

$$\left| \begin{matrix} u \\ v \end{matrix} \right| = \begin{bmatrix} N1 & 0 & N2 & 0 & N3 & 0 & N4 & 0 \\ 0 & N1 & 0 & N2 & 0 & N3 & 0 & N4 \end{bmatrix} \begin{bmatrix} u1 \\ vi \\ u2 \\ v2 \\ u3 \\ v3 \\ u4 \\ v4 \end{bmatrix} \tag{10}$$

Where,

u1, v1….u4, v4 are nodal displacement and *N1-N4* are shape function and that are connected with the nodes 1–4 correspondingly.

6.3 Coordinate transformation

The shape function is additionally used for coordinate's alteration of element in order to simplify the integration for calculation of stiffness matrix of some quantities for element. The coordinates (x, y, z), within the element of a point represented by Eq. (11) [28].

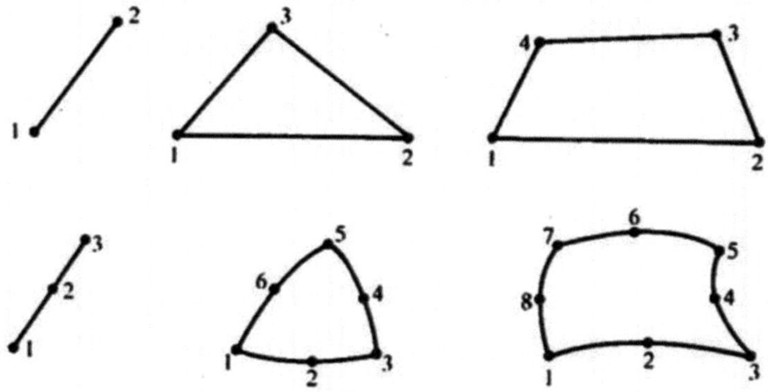

Figure 8.
Some element forms and node position used in two dimensional [28].

$$x = \sum_{i-1}^{n} N_i x_i$$
$$y = \sum_{i-1}^{n} N_i y_i \quad (11)$$
$$z = \sum_{i-1}^{n} N_i z_i$$

6.4 Relation between strain and displacement

For two dimensional element domain the relation between strain and displacement represent by Eq. (12) [28].

$$\varepsilon = \begin{bmatrix} \varepsilon_x \\ \varepsilon_y \\ \gamma_{xy} \\ \varepsilon_z \end{bmatrix} = B \begin{bmatrix} u1 \\ v1 \\ u2 \\ v2 \\ \dots \\ un \\ vn \end{bmatrix} \quad (12)$$

6.5 Relation between stress and strain

It may express as:

$$\Delta\sigma = D_T \Delta\varepsilon \quad (13)$$

Where,

$\Delta\sigma$ is the vector of stress components, $\Delta\varepsilon$ represents corresponding components of strains and DT is a square matrix that is constant in the elastic case.

6.6 Global stiffness matrix

It is formed when added the stiffness matrices of all elements. The equation for global stiffness is given as:

$$K\Delta\delta = \Delta R \quad (14)$$

Where.

$\Delta\delta$ is unknown vector having increments of nodal displacement due to increment force ΔR.

For material linear elastic material behavior the equation may by write as (Scheldt, 2002).

$$K\delta = R \quad (15)$$

6.7 Finite element based software's

Following are finite element based software's.

1. Displacement Analyzers Finite Element program (DIANA) software is developed by TNO Building and Construction Research, Netherlands. It is a flexible software and used in solving of linear and nonlinear structural engineering in 2D and 3D [28].

2. Phase2 developed by rock science for solving 2D non-linear problems like analysis of displacements and stresses around underground openings, in the field of mining and civil engineering [29].

3. ABAQUS software is developed by Hibbitt et al. (1978) in USA. It is used for linear and non-linear, problems and analyzes the stresses of any structure in 3D [28].

4. ANSYS software is developed for solving both linear and non-linear problems for isotropic and non-isotropic properties of materials [28].

7. Conclusion

Engineering design is the valuation using knowledge of basic sciences, mathematics and engineering sciences to convert resources optimally to meet quantified objectives. Its goal is to develop a solution to a known problem. There are different stages of design process; one can adapt it to the particular problem for solving it. We have variety of design techniques in rock engineering. They are classified in to three groups i.e. are Analytical, Empirical and Observational. Among, these empirical approaches can effectively be used for engineering judgment. Rock mass classification is one of the widely used empirical methods for the assessment of the rock mass behavior. The empirical methods of design do not estimate accurately the reliability of support systems, redistribution of stresses and rock mass deformation. Numerical methods are considered very useful to be used for estimate these parameters precisely and in short time as compared to other methods of design. So it is recommended that for efficient and viable design the numerical and empirical methods should be used in parallel for the assessment of soil/rock mass behavior to design any underground structure.

Acknowledgements

We acknowledge the support of all colleagues in the Department of Mining Engineering, University of Engineering and Technology Peshawar, Pakistan while compiling this work.

Conflict of interest

We have no conflict of interest.

Notes/thanks/other declarations

Thanks and warm regards.

Author details

Zahid Ur Rehman[1]*, Sajjad Hussain[1], Noor Mohammad[1], Akhtar Gul[2] and Bushra Nawaz[1]

1 Department of Mining Engineering, University of Engineering and Technology Peshawar, Pakistan

2 Department of Civil Engineering, University of Engineering and Technology Peshawar, Pakistan

*Address all correspondence to: engr.zahid@uetpeshawar.edu.pk

References

[1] https://www.google.com.pk/search?q=optimization&oq=optimization&aqs=chrome..69i57j69i59j69i60l3j69i61.2975j0j7&sourceid=chrome&es_sm=93&ie=UTF-8

[2] Barton, N., Lien, R., & Lunde, J. (1974). Engineering classification of rock masses for design of rock support. *Rock Mechanics, 189–236.*

[3] Bieniawski, Z. (1984). *Rock Mechanics Design in Mining and Tunneling* . Rotterdam, Netherlands: A. A.Balkema.

[4] Stacey, T. (2004). The link between the design process in rock engineering and the code of practice to combat rock fall and rockburst accidents. *The Journal of The South African Institute of Mining and Metallurgy*, 29–34.

[5] Khandani, S. (2005). *Engineering Design Process.* Saylor.org.

[6] Bieniawski, Z. T. (1990). *Tunnel Design By Rock Mass classifications.* Washington: Department of the army, US army Corpss of Engineer Washington, DC 20314–1000.

[7] Ali, E. W. (2014). *M.Sc thesis, rock mass charecterization for diversion tunnel at diamer basha dam, pakistan-a design perspective.* peshawar: University of Engineering and Technology, Peshawar, Pakistan.

[8] *www. google.com.* (2016, March Thursday). Retrieved 2016 Thursday, 2016, from www. google.com: https://www.rocscience.com/documents/hoek/corner/04_Rock_mass_classification.pdf

[9] Muhammad Tahir. (2014). Prediction performance and Generalization of the emeprical Estimation of Rock mass Deformation Modulus Based on Rockmass Classification Systems. *International Journal of Scientific Engineering and Technology*, 1488–1495.

[10] *www.google.com*. (2015). Retrieved from www.google.com

[11] Karahan, E. (2010). *Design of Excavation and Support System for the Cubukbili Tunnel in Antalya.*

[12] E.Hoek, P. K. (1993). *Support of Underground Excavations in Hard Rock.*

[13] E.HOEK, P. K. (2016, March Wednesday). *www.google.com*. Retrieved March 2016, from www.google.com: http://web.mst.edu/~rogersda/umrcourses/ge341/Rock%20Mass%20Rating.pdf.

[14] (NGI), N. g. (2013). *Using the Q-System.* Sweden and Norway: NGI.

[15] D. Milne, J. H. (1998). Rock mass characterization for underground hard rock mines. *Tunnelling and Underground Space Technology*, 383–391.

[16] Pantelidis, L. (2009). Rock Slope Stability assestment through rock mass calsification systems. *International Journal of Rock Mechanics and Mining Science*, 315–325.

[17] Sonmez, H., & Ulusay, R. (1999). Modifications to the geological strength index (GSI) and their applicability to stability of slopes. *Int J Rock Mech Min Sci., 36*, 763–760.

[18] V. Marinos, P. E. (2005). Gelogical strength index: applications and

[19] Mahmoud Hashemi, S. M. (2010). Application of rock mass charecterization for determining the mechanical properties of rock mass: a comparitive study. *Rock Mechanics, Rock Engineering*, 305–320

[20] H. Basarir, A. O. (2005). Analysis of support requirements for a shallow

diversion tunnel at Guledar dam site, turkey. *Engineering Geology*, 131-[

[21] Bobet, A. (2010). Numerical methods in geomechanics. *The Arabian Journal for Science and Engineering, 35*, Number 1B.

[22] Zulfu Gurocak, P. S. (2007). Empirical and numerical analyses of support requirements for a diversion tunnel at the Boztepe dam site, eastern Turkey. *Engineering Geology, 91*, 194–208.

[23] M. Genis, H. B. (2007). Engineering geological appraisal of the rock masse s and preliminary support design, Dorukhan Tunnel, Zonguldak, Turkey. *Engineering Geology, 92*, 14–26.

[24] Aydin Ozsan, H. B. (2006). Engineering geological investigations along the Ankara subway extension. *IAEG*, paper no. 586.

[25] Rasouli, M. (2009). Engineering geological studies of the diversion t

[26] John A. Hudson, X.-T. F. (2010). Technical auditing of rock mechanics modelling and rock engineering design. *International Journal of Rock Mechanics & Mining Sciences, 47*, 877–886.

[27] L. Jing, J. H. (2002). Numerical methods in rock mechanics. *International Journal of Rock Mechanics & Mining Sciences, 39*, 409–427.

[28] Jing, L. (2013). A reviewof techniques, advances and outstanding issues in numerical modelling for rock mechanics and rock engineering. *International Journal of Rock Mechanics & Mining Sciences, 40*, 283–353.

[29] M. Caia, P. K. (2004). Estimation of rock mass deformati on modulus and strength of jointed hard rock masses using the GSI system. *International Journal of Rock Mechanics & Mining Sciences, 41*, 3–19.

Chapter 4

Three Dimensional Slope Stability Analysis of Open Pit Mine

Masagus Ahmad Azizi, Irfan Marwanza, Muhammad Kemal Ghifari and Afiat Anugrahadi

Abstract

The 3-dimensional slope stability analysis has been developing rapidly since the last decade, and currently a number of geomechanical researchers in the world have put forward ideas for optimization of slope design related to the economics and safety of mining operations. The 3-dimensional slope stability analysis methods has answered the assumption of spatial parameters in determining safety factors and the failure probability, thus the volume of failed material and the location of the most critical slopes can be determined. This chapter discusses two methods of 3-dimensional slope stability analysis, namely the limit equilibrium method (LEM) and finite element method (FEM). LEM 3D requires an assumption of failure type with the variable of analysis are the maximum number of columns, the amount of grid points, increment radius, and type of slip surface. On the other hand, FEM 3D requires an assumption of convergence type, absolute force and energy, with the variable of analysis are mesh type and maximum number of iterations. LEM 3D shows that the cuckoo algorithm is reliable in obtaining position and shape of slip surface. Meanwhile FEM 3D, the optimum iteration number needs to be considered to improve analysis efficiency and preserving accuracy.

Keywords: open pit mine, slope stability analysis, 3D LEM and FEM

1. Introduction

This chapter will mainly discuss about 3D slope stability analysis using Limit Equilibrium Method (LEM) and Finite Element Method (FEM). These two methods are widely developed by academics and have been applied by many mining geotechnical practitioners in slope stability analysis. In recent time most of the analysis are performed in 2D method because of its simplicity and lower operational cost [1]. However, 3D analysis is more justifiable to represent the actual geometry condition. Thus, to obtain 3D slope stability analysis has its own importance. In this regard, 2D and 3D analysis can both be performed for the same slope stability analysis problem to obtain a more convincing and realistic results. These two methods can be used to validate one to another [2]. The analysis of the 3D approach is carried out by [3–5] in the 1970's. Ref. [5] modifying the slice model used in 2D analysis to become a column in 3D analysis. The study of 3D slope stability analysis model is further developed and evaluated by [2, 6, 7] states that 3D FoS value is always greater than

IntechOpen

2D analysis, therefore 2D results are considered more conservative [8]. Most existing three-dimensional 3D slope stability analysis methods are based on simple extensions of corresponding two-dimensional 2D methods of analysis and a plane of symmetry or direction of slide is implicitly assumed. 3D asymmetric slope stability models based on extensions of Bishop's simplified, Janbu's simplified, and Morgenstern–Price's methods are developed by [8].

2. Slope stability analysis

The stability of a slope can be determined by 2 criteria of considerations, which is the value of the safety factor (FoS) and also the value of the probability of failure (PoF), these two criteria are used to determine the optimal geometry of the pit opening. In order to obtain accurate analysis results, the information data regarding the geotechnical conditions of the model must be repetitive to the actual conditions. In general, the principal of the value of safety factor concept is the ratio between the shear strength along the slip surface required to maintain the slope at a stable condition, and the available shear strength [9]. The above definition can be mathematically described as:

$$\mathrm{FoS} = \frac{S_u}{\tau_{\mathrm{required}}} \tag{1}$$

The slope is assumed to be a model of an inclined plane [9]. By determining the resultant overall forces and moments acting in equilibrium state, the slope factor of safety value can be determined by comparing the amount of the resisting force to the driving force, or the resisting moment to overturning moment can see in **Figure 1**.

$$\mathrm{FoS} = \frac{\text{Resisting force}}{\text{Mobilized force}} \tag{2}$$

$$\mathrm{FoS} = \frac{\text{Resisting moment}}{\text{Overturning moment}} \tag{3}$$

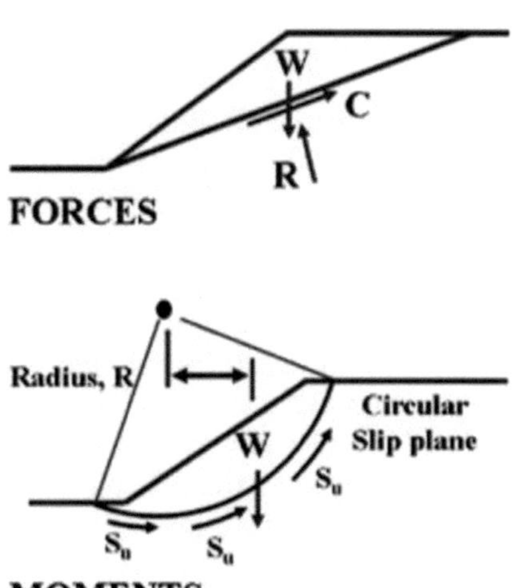

Figure 1.
Equilibrium force and moment in inclined plane [9].

Based on the Mohr-Coulomb failure criteria, the definition of the value of the safety factor can be determined as follows:

$$\mathrm{FoS} = \frac{c + \sigma \tan\phi}{\tau_{\mathrm{required}}} \tag{4}$$

Where c and ϕ are effective cohesion and internal friction angle.

The probabilistic failure is an approach that consider various input parameters that generates different Factor of Safety (FoS) values [10]. This information is based on the fact that every random input parameter has the same probability to yield a certain value of FoS. Regarding the difficulty and high expense of field and laboratory data collecting, this method is more attractive because of its representativeness. **Figure 2** presents the concept of failure probability and the amount of uncertainty. Slope PoF is determined from the ratio between the area under the curve of the distribution of FoS <1 value to the distribution of FoS ≥1 value. The greater the range of distribution of FoS values, the higher the uncertainty of FoS values with the same PoF values.

By definition there is a linear relationship between the PoF value and the likelihood of failure, while this does not apply to the FoS relationship with the chance of failure [10]. A large FoS does not represent a more stable slope, because the implicit uncertainty is not captured by the FoS value. Slopes with FoS of 3 do not mean that they are 2 times more stable than FoS of 1.5, while slopes with a PoF value of 5% are 2 times more stable than slopes with a PoF value of 10%. Slope stability in general performed in two-dimensional analysis. But in modeling complex geometries, 2D analysis cannot simulate them properly. Therefore, the 3D analysis is considered to be able to describe the conditions in the field better than the 2D analysis. Analysis of slope stability with 3D limit equilibrium method starts by assuming the geometry of the sliding mass (**Figure 3**).

The results of the calculation of slope stability can be expressed in safety factor. In this method, safety factor is not only influenced by the direction sliding, but also by the slip surface that safety factor is sensitive to critical slip surface locations. Therefore, the determination of a critical slip surface is very important. Safety factor can be obtained correctly if the determination of critical slip surface is accurate.

2.1 Slope design

Optimal slope geometry is obtained from a step-by-step assessment process [12] state there are 5 stages of the process, which is models, domains, design, analysis

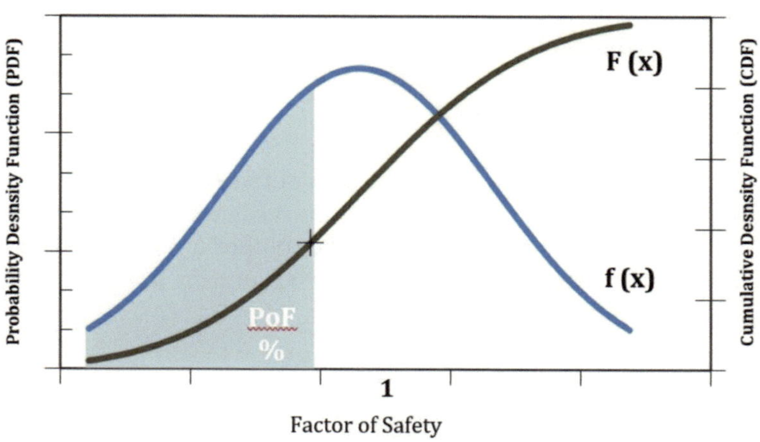

Figure 2.
Concept of probabilistic failure.

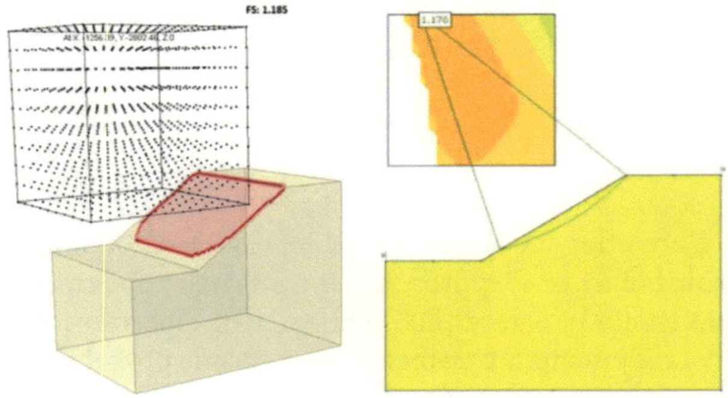

Figure 3.
Comparison between 3D and 2D single slope analysis [11].

and implementation. The initial stage of the geotechnical model is determined by 4 parameters, namely the geological model, structural conditions, rock mass and hydrogeological model. At the domains stage, the failure modes are determined by 2 parameters, namely the strength of the material and the condition of the structure. A single slope design arrangement is determined by the regulations or standards used by the company and the capabilities of the equipment used. Determination of the haulage road width and also the overall slope angle is based on mine planning related to the economic aspects of the opening geometry made. Furthermore, the stability analysis of the slope geometry that has been designed refers to the parameters (structure, strength, groundwater, and in-situ stress). After the final design is obtained, a risk assessment is carried out to mitigate the potential for landslides that may occur. In the implementation stage, the functions of dewatering, blasting and monitoring of the progress of the design model and the movement of rock masses (**Figure 4**).

2.2 Limit equilibrium method 3D

These days the needs and pressure to analyze a slope 3 dimensionally is more sounds. This is because 2D analysis assumes that the width of slope is infinitely wide so then it neglects 3D effect [11]. In most of the cases the width to height ratio is not sufficiently long and varies perpendicular to the slide movement. Therefore, 3D analysis is considered important to be done to produce the representative FoS. Moreover, in 3D analysis the volume of failure can also be estimated while 2D analysis cannot. If the volume can be determinate, it can be useful as one of the considerations in giving failure prevention recommendation.

2.2.1 Safety factor for 3D slope stability

The 3-dimensional model is a refined version of the 2-dimensional by projecting the skid plane into a column and determining the resultant force, as well as the moment based on the x, y, and z directions. The equilibrium force and moments acting on the overall column mass are used to determine the following 3 possible direction of the slip plane:

1. The column moves in the same direction

2. The column moves towards one another

3. The column moves in the opposite direction

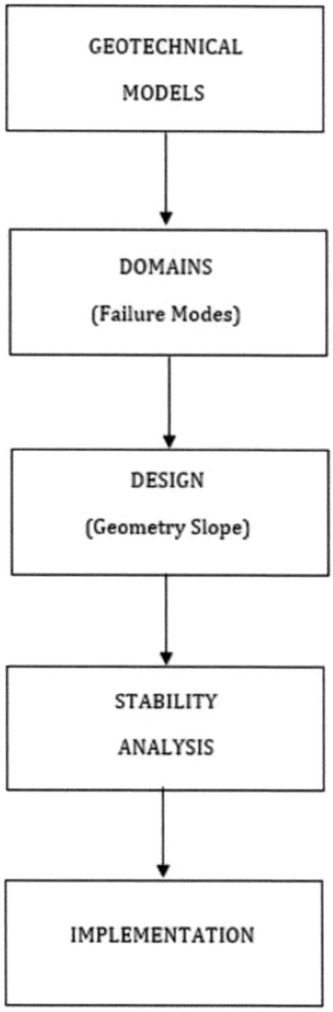

Figure 4.
Slope design process [12].

For the 3-dimensional analysis, the mass potential of the slip plane is divided into several columns. Ref. [8] give the equation of the Simplified Janbu method deduced from the Morgenstern-Price method to obtain a safety factor value of 3-dimensional analysis (**Figure 5**).

a_i is space angle for sliding direction with respect to the projected $x - y$ plane, a_x, a_y are base inclination along x and y directions measure at the center of each column, E_{xi}, E_{yi} are inter-column normal forces in x and y directions, respectively, H_{xi}, H_{yi} are lateral inter-column shear forces in x and y directions, N'_i, U_i are effective normal and base pore watery force, P_{vi}, S_i is vertical external force, and base mobilized shear force, and X_{xi}, X_{yi} are vertical inter-column shear force in plane perpendicular to x and y directions.

With the mohr-coulomb collapse criteria, the safety factor is determined using the following equation:

$$F = \frac{S_{fi}}{S_i} = \frac{C_i + N'_i \tan \phi'_i}{S_i} \quad (5)$$

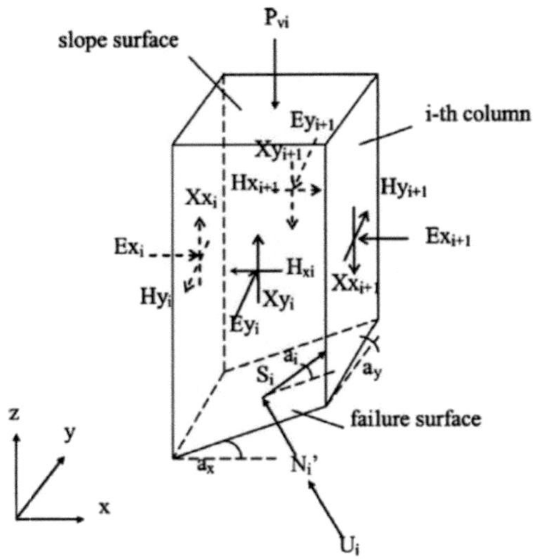

Figure 5.
3 dimensional column [8].

where S_{fi} is ultimate resultant shear force available at the base of column i, N'_i is the effective base normal force, C_i is (c. Ai) and c and A_i are effective cohesive strength and the base area of the column. The base shear force S_i and normal base force N_i are expressed as the components of forces with respect to x, y, and z directions for column i.

$$S_{xi} = f_1 S_i, S_{yi} = f_2 S_i, S_{zi} = f_3 S_i \tag{6}$$

$$N_{xi} = g_1 N_i, N_{yi} = g_2 N_i, N_{zi} = g_3 N_i \tag{7}$$

where f_1, f_2, f_3 and g_1, g_2, g_3 = unit vectors in the direction of S_i and N_i. The projected shear angles a' = same for all columns in the x – y plane in the present formulation, and by using this angle, the space shear angle ai found for each column.

$$a_i = \tan^{-1}\{\sin\theta_i / [\cos\theta_i + (\cos a_{yi} / \tan a' \cos a_{xi})]\} \tag{8}$$

$$\theta_i = \cos^{-1}\{\sin a_{xi} \bullet \sin a_{yi}\} \tag{9}$$

An arbitrary intercolumn shear force function f (x, y) is assumed in the present analysis, and the relationships between the intercolumn shear and normal forces in the x- and y-directions are given as:

$$Xx_i = Ex_i\, f(x,y)\lambda_x \tag{10}$$

$$Xy_i = Ey_i\, f(x,y)\lambda_y \tag{11}$$

$$Hx_i = Ey_i\, f(x,y)\lambda_{xy} \tag{12}$$

$$Hy_i = Ex_i\, f(x,y)\lambda_{yx} \tag{13}$$

Where λ_x and λ_y are intercolumn shear force mobilization factors in x and y directions, respectively, and λ_{xy} and λ_{yx} are intercolumn shear force mobilization factors in xz and yz planes. Considering the vertical and horizontal force equilibrium for the ith column in the z, x, and y directions produces the following equations (**Figure 6**):

$$F_z = 0 = N_i g_{3i} + S_i f_{3i} - (W_i + P_{vi}) = (X_{xi+1} - X_{xi}) + (X_{yi+1} - X_{yi}) \tag{14}$$

$$F_x = 0 = S_i f_{1i} - N_i g_{1i} - H_{xi} + H_{xi+1} = E_{xi+1} - E_{xi} \tag{15}$$

$$F_y = 0 = S_i f_{2i} - N_i g_{2i} - H_{yi} + H_{yi+1} = E_{yi+1} - E_{yi} \tag{16}$$

Solving Equation, the base normal and shear forces can be expressed as

$$N_i = A_i + B_i S_i \tag{17}$$

$$S_i = \frac{C_i + (A_i - U_i)\tan\phi'_i}{F\left[1 - \left(\frac{B_i \tan\phi_i}{F}\right)\right]} \tag{18}$$

$$A_i = \frac{W_i + (P_{vi} + \Delta Ex_i \lambda_x + \Delta Ey_i \lambda_y)}{g_{3i}} \tag{19}$$

$$B_i = -\frac{f_{3i}}{g_{3i}} \tag{20}$$

$$F = \frac{C_i + N'_i \tan\phi'_i}{S_i} \tag{21}$$

Considering the overall force equilibrium in x-direction internal force E cancels out.

$$-\sum H_{xi} + \sum N_i g_{1i} - \sum S_i f_{1i} = 0 \tag{22}$$

Considering overall moment equilibrium in the x-direction

$$-\sum (W_i + P_{vi} - N_i g_{3i} - S_i f_{3i}) RX + \sum (N_i g_{1i} - S_i f_{1i}) RZ = 0 \tag{23}$$

Considering overall force equilibrium in the y-direction

$$-\sum H_{yi} + \sum N_i g_{2i} - \sum S_i f_{2i} = 0 \tag{24}$$

Considering overall moment equilibrium in the y-direction

$$-\sum (W_i + P_{vi} - N_i g_{3i} - S_i f_{3i}) RY + \sum (N_i g_{2i} - S_i f_{2i}) RZ = 0 \tag{25}$$

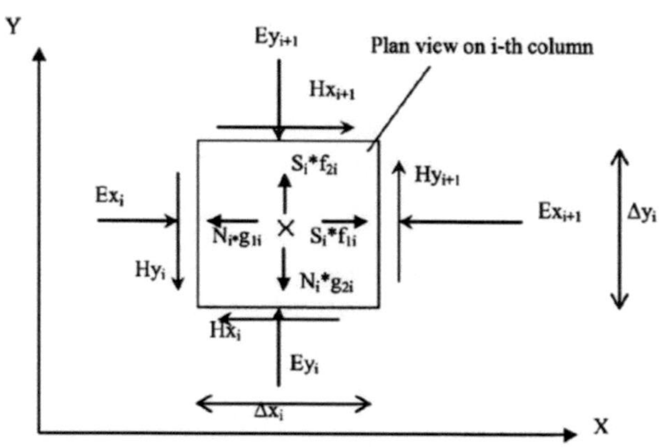

Figure 6.
Force equilibrium in columns [8].

The directional safety factor F_x and F_y is determined as follows:

$$F_x = \frac{\sum[C_i + (N_i - U_i)\tan\phi_i]f_{1i}}{\sum N_i g_{1i} - \sum H_{xi}}, \quad (0 < F_x < \infty) \tag{26}$$

$$F_y = \frac{\sum[C_i + (N_i\_U_i)\tan\phi'_i]f_{2i}}{\sum N_i g_{2i} - \sum H_{yi}}, \quad (0 < F_y < \infty) \tag{27}$$

Formulation 3D Bishop's methods by considering the overall moment equilibrium equations in x or y direction and neglecting the inter-column vertical and horizontal shear forces.

$$F_{mx} = \frac{\sum\{K_{xi}[f_{2i}RZ_i + f_{3i}RY_i]\}}{\sum(W_i + P_{vi})\,RY_i + \sum N_i(g_{2i}RZ_i - (g_{3i}RY_i)} \tag{28}$$

$$F_{my} = \frac{\sum\{K_{yi}[f_{1i}RZ_i + f_{3i}RX_i]\}}{\sum(W_i + P_{vi})\,RX_i + \sum N_i(g_{1i}RZ_i - (g_{3i}RX_i)} \tag{29}$$

$$K_{xi} = \left\{\frac{C_i + [(W_i + P_{vi})/(g_{3i} - U_i)]\tan\phi_i\,\}}{1 + (f_{3i}\tan\phi_i/g_{3i}\,F_{mx})}\right. \tag{30}$$

$$K_{yi} = \left\{\frac{C_i + [(W_i + P_{vi})/(g_{3i} - U_i)]\tan\phi_i\,\}}{1 + (f_{3i}\tan\phi_i/g_{3i}\,F_{my})}\right. \tag{31}$$

Considering overall moment equilibrium about an axis passing through (x_0, y_0, z_0) and parallel to the z axis gives:

$$\sum(-N_i g_{1i} - S_i f_{1i})RY + \sum(N_i g_{2i} - S_i f_{2i})RX = 0 \tag{32}$$

$$F_{mz} = \frac{\sum[K_{zi}(f_{2i}RX_i - f_{3i}RY_i)]}{\sum N(g_{2i}RX_i - (g_{1i}RY_i)} \tag{33}$$

$$K_{zi} = \left\{\frac{C_i + [(W_i + P_{vi})/(g_{3i} - U_i)]\tan\phi_i\,\}}{1 + (f_{3i}\tan\phi_i/g_{3i}\,F_{mz})}\right. \tag{34}$$

For the 3D asymmetric Bishop's method, at moment equilibrium point, the directional factors of safety, F_{mx}, F_{my}, and F_{mz} are equal to each other. Under this condition, the global factor of safety F_m based on moment can be determined as

$$F_m = F_{mx} = F_{my} = F_{mz} \tag{35}$$

Formulation 3D simplified Janbu's methods by considering the overall force equilibrium equations and neglecting the inter-column vertical and horizontal shear forces.

$$A_{xi} = \frac{\{C_i + [(W_i\_P_{vi})/(g_{3i}\_U_i)]\tan\phi_i\}}{1 + (f_{3i}\tan\phi_i/g_{3i}F_{sx})} \tag{36}$$

$$A_{yi} = \frac{\{C_i + [(W_i\_P_{vi})/(g_{3i}\_U_i)]\tan\phi_i\}}{1 + (f_{3i}\tan\phi_i/g_{3i}F_{sy})} \tag{37}$$

$$F_{sx} = \frac{\sum A_{xi}(f_{1i} + f_{3i}g_{1i}/g_{3i})}{\sum(g_{1i}/g_{3i})(W_i + P_{vi})} \tag{38}$$

$$F_{sy} = \frac{\sum A_{yi}\left(f_{2i} + f_{3i}g_{2i}/g_{3i}\right)}{\sum\left(g_{2i}/g_{3i}\right)\left(W_i + P_{vi}\right)} \tag{39}$$

For 3D asymmetric Janbu's method, at the force equilibrium point, the directional factors of safety, F_{sx}, and F_{sy} are equal to each other. Under this condition, the global factor of safety F_f based on force is determined as follows:

$$F_f = F_{sx} = F_{sy} \tag{40}$$

The safety factor is also used in vertical and 3D force equilibrium to achieve the simplified Janbu's method.

2.2.2 Grid search in determination of slip surface

One of the methods that can be used to determine the critical slip surface is the grid search method [11]. In the grid search method, the first thing to do is to determine the size of the grid box with dimensions x, y, and z. After the grid box is available, the user determines the number of grid points that you want to use in the x, y, and z directions. This point serves as the center of rotation. Each center of rotation can produce a number of circles that are used as slip surfaces. The number of circles produced at each center of rotation from the minimum radius to the maximum radius is called the radius increment. Illustration of the number of grid points and radius increment can be seen in **Figures** 7 and **8**.

After assuming the field geomaterial failure, the next step is mass discretization of the sliding mass into a number of columns. Square nets are applied to the sliding mass so that the sliding mass is divided into columns. There are two kinds of columns; the active column where the column is inside the sliding mass boundary line, and the inactive column where these columns are outside the sliding mass boundary line. In the calculation, the inactive columns are ignored so that the discrete sliding mass is determined only from the sum of the active columns. **Figure 9** shows the illustration of the discretization of the sliding mass using a square grid.

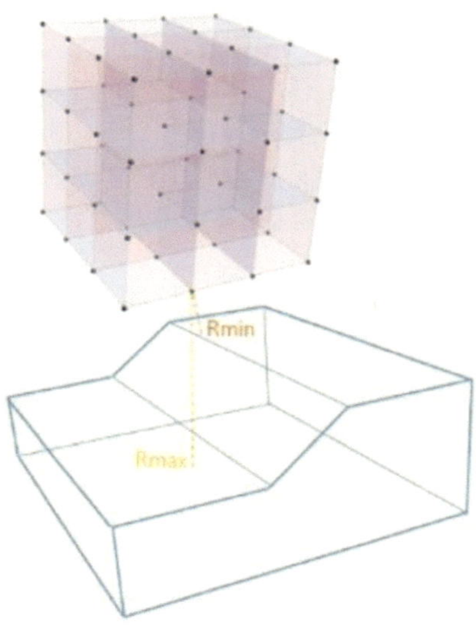

Figure 7.
Illustration of grid point [11].

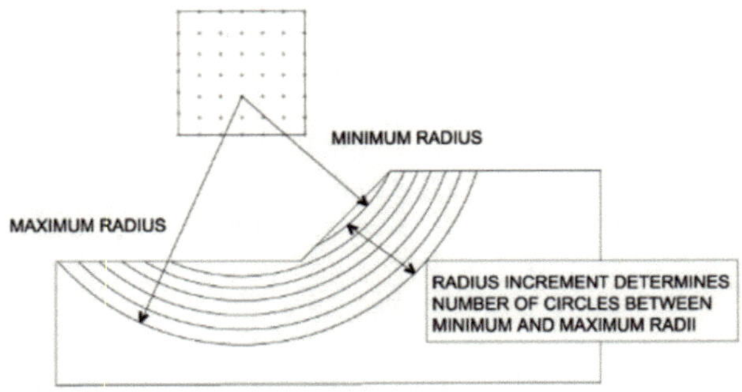

Figure 8.
Illustration of the radius increment in the grid search [11].

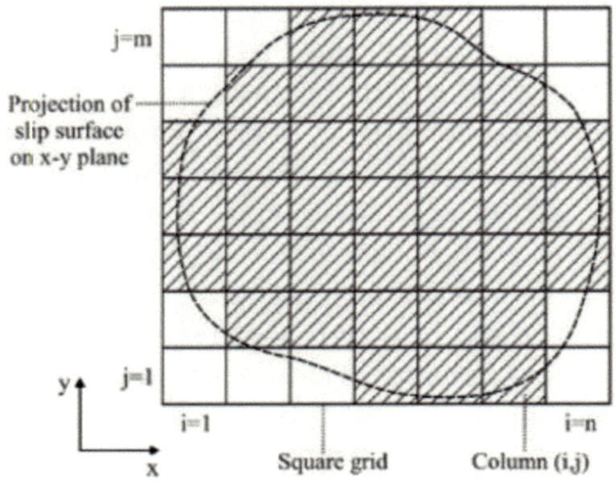

Figure 9.
Discretization of the sliding mass using a square grid [11].

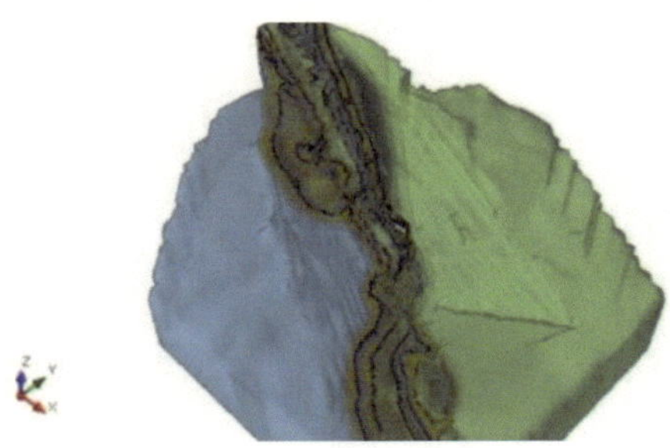

Figure 10.
3D model for slope stability analysis [11].

After discretizing the sliding mass, internal and external forces in each column can be calculated based on moment equilibrium, force equilibrium, or both depending on what method of calculation is used (**Figure 10**).

The grid search method is used to find critical slip surface. The grid search method starts by specifying the grid box dimension. The location and dimensions of the grid box must cover the entire study area so that the search for critical slip surface can be performed optimally, for the influence of number of grid point and radius increment in determining safety factor result can be seen in **Table 1**. The result of 3D slope stability analysis using grid search can see in **Figure 11**.

Radius increment	Number of grid points	Factor of safety	Volume (m^3)	Location	Direction of sliding	Center of rotation		
						X	Y	Z
	20 × 20 × 10	0.868	2,245,470	North-HW	246.7	371,070	9,586,190	470
10	30 × 30 × 15	0.874	1,998,950	North-HW	246.6	371,069	9,586,190	446
	40 × 40 × 20	0.873	2,009,660	North-HW	246.5	371,068	9,586,190	443
	20 × 20 × 10	0.868	2,245,470	North-HW	246.7	371,070	9,586,190	470
20	30 × 30 × 15	0.874	2,028,600	North-HW	246.5	371,069	9,586,200	429
	40 × 40 × 20	0.872	1,934,080	North-HW	246.4	371,069	9,586,200	428
	20 × 20 × 10	0.868	2,245,470	North-HW	246.7	371,070	9,586,190	470
30	30 × 30 × 15	0.906	1,628,780	North-HW	246.2	371,092	9,586,210	352
	40 × 40 × 20	0.867	1,937,390	North-HW	246.1	371,076	9,586,170	459
	20 × 20 × 10	0.868	2,245,470	North-HW	246.7	371,070	9,586,190	470
40	30 × 30 × 15	0.876	1,756,790	North-HW	246.2	371,076	9,586,200	404
	40 × 40 × 20	0.872	1,934,080	North-HW	246.4	371,069	9,586,200	428
	20 × 20 × 10	0.868	2,245,470	North-HW	246.7	371,070	9,586,190	470
50	30 × 30 × 15	0.876	2,068,950	North-HW	246.3	371,077	9,586,180	459
	40 × 40 × 20	0.879	1,968,990	North-HW	246.3	371,078	9,586,190	421

Table 1.
The influence of number of grid point and radius increment in determining safety factor.

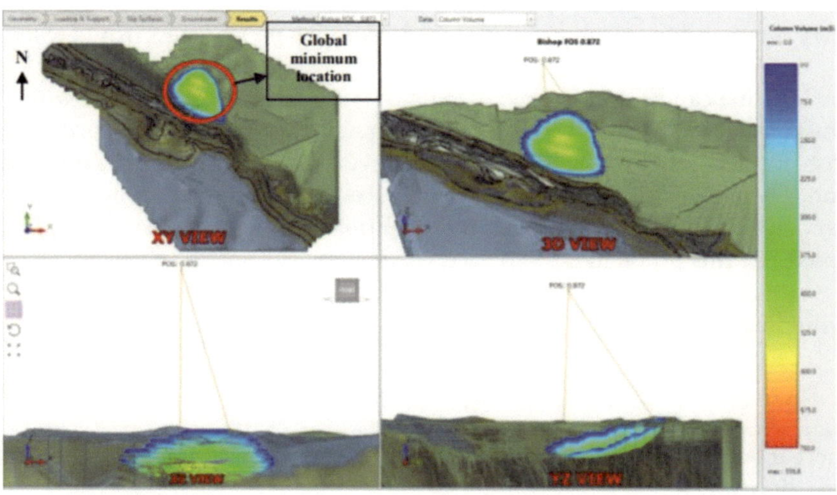

Figure 11.
Grid search LEM 3D analysis result [11].

2.2.3 Cuckoo search in determination of slip surface

Grid Search is commonly used as slip surface searching method because the principle is simple and easy to understand [11]. However, this method can only calculate the circular slip surfaces, so it cannot represent the stability of slope in real condition. To make it more representative, non-circular slip surface option is also available in slope stability simulation software, and one of the searching methods in non-circular option is CS method.

CS is an algorithm which is used for solving the optimization problems. CS has been used in engineering field, such as welded beam and spring design optimization but there is still a few that use it for slope stability issue. Nowadays, the necessity to analyze 3D slope stability is more essential. The reason why slope stability problem should not be assumed 2 dimensionally that should be taken into account is the importance of determining volume of failure for risk management volume of failure is counted as one of the consequences, and it can be obtained by analyzing slope stability in 3D, indeed. Furthermore, there is a relationship between failure probability and volume of failure. Another issue exists when 3D analysis is performed using Grid Search on a vast area. The use of 3D analysis on a vast area can be complicated, and in real cases more advanced optimization methods are required. Therefore, CS is tried to be applied in order to determine the slip surface with minimum SF in 3D analysis. Thus, it can be suggested to be used as an alternative or other better option in 3D analysis.

CS means a metaheuristic optimization method that was developed by [13]. This method was inspired from cuckoo's breeding behavior. In this research, CS is used as a slip surface search tool that has the lowest SF. CS is coupled with Lévy Flights random walk. There are few rules to use this algorithm as follows:

- Each cuckoo lay one egg at a time, and dumps it in a randomly chosen nest;

- The best nests with high quality of eggs (solutions) will carry over to the next generations;

- The number of available host nests is fixed, and a host can discover an alien egg with a probability pa $\in [0, 1]$. In this case, the host bird can either throw the egg away or abandon the nest so as to build a completely new nest in a new location.

The last rule can be approached using a fraction pa to determine the worst solutions of n nests that will be replaced with a new nest randomly. In order to solve the problem, it can be simply illustrated that every egg in a nest represents one new solution. The purpose is to use the new and potentially better solution to replace the current solution in the nest. In a certain condition, the nest may have 2 eggs (2 solutions) but this problem is simplified so one nest has only 1 solution.

The random walk as determinated by Lévy Flights can be described in the following formula:

$$x_i^{(t+1)} = x_i^{(t)} + \alpha \bigoplus \text{Lévy}(\lambda) \quad (41)$$

where $\alpha > 0$ is the step size and Lévy(λ) is the position function from Lévy Flights (**Figure 12**).

$$\text{Lévy} \sim u = t^{-\lambda}, (1 < \lambda \leq 3) \quad (42)$$

CS has been applied in many optimizations and computer intelligence with promising efficiency, this has been proved from design application in engineering field, scheduling problems, thermodynamic calculations, etc. Few examples of CS application in engineering field are designing spring, welded beam, and steel frame. The CS's performance has also been compared with some metaheuristic algorithms such as PSO and GA, and the result shows that CS has higher success rate than. The result of the influence of max columns in x or y and max iteration in determining safety factor can be seen in **Table 2**.

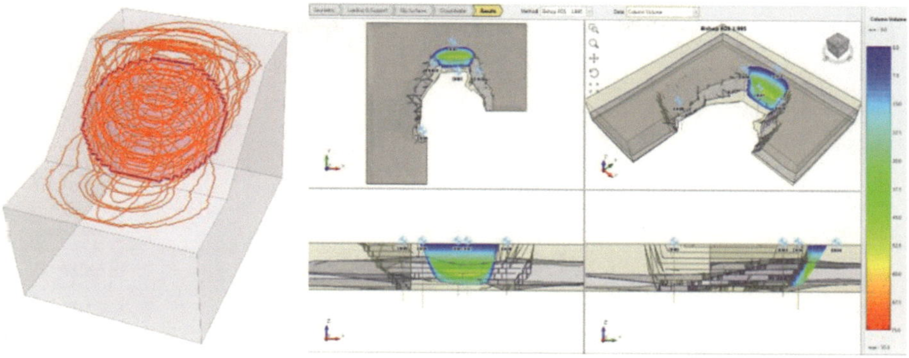

Figure 12.
Cuckoo search LEM 3D analysis result [11].

2.2.4 Analysis 3D limit equilibrium of open pit mine

3D limit equilibrium analysis method, data regarding 3D slope geometry model, material properties and 3D geological models are required. The 3D slope geometry model required for this method can be obtain from the reconstruction of pit surface model from either terrestrial or photogrammetric measurement methods. The next step is to create an external volume from the pit surface model which will then be analyzed to determine the position and shape of the slip surface, example of 3D slope geometry model can be seen in **Figure 13**.

Limit equilibrium analysis method uses the properties values of materials obtained from laboratory tests results to calculate the value of the safety factor. An example of the input parameter data used in the analysis can be seen in **Table 3**.

Geological modeling is the process of creating visual description geometry of rock lithology into software that represents the actual conditions. However, there are limitations in the modeling process, that related to the limited information on the data held and errors in data interpretation carried out. One of the methods that can be used to create the 3D geological model is interpolation the lithology information data from geotechnical drilling results. The 3D geological model will be used in the 3-dimensional slope stability analysis to determine the distribution characteristics of the rock lithology. In this analysis, the 3D geological floor model of limonite, saprolite and bedrock lithology is used and can be seen in **Figure 14**.

The results of the analysis using the Bishop Simplified method for the slip surface search method, both grid search and cuckoo search can be seen in **Figures 15** and **16**. The analysis results with the grid search show the safety factor value is 1.104, while the cuckoo search is 1.089. The position of the slip surface with the grid search and cuckoo search is the same position it indicates the accuracy of the cuckoo search, while the grid requires a grid box which must represent the actual slip surface position.

2.3 Finite element method 3D

The finite element method has been widely applied by mining geotechnical practitioners in slope stability analysis, with the advantage that the stress–strain analysis in the material allows to determine the displacement and strain values acting on the model elements, but this method has weaknesses in the process. This analysis uses a large number of calculation matrices so that it requires a long computation time, especially if the analysis is carried out in 3D, the number of

Max iteration	Max columns in X or Y	FoS	Volume (m^3)
40	50	2.06	159,816
	60	2.02	179,523
	70	2.10	229,704
	80	2.10	303,186
	90	2.01	183,631
	100	2.01	187,898
80	50	2.03	179,192
	60	2.02	192,951
	70	2.03	167,961
	80	2.02	178,858
	90	2.02	171,821
	100	2.01	183,469
120	50	2.02	175,113
	60	2.02	180,573
	70	2.01	178,671
	80	2.01	193,606
	90	2.01	180,490
	100	2.02	190,336
160	50	2.00	194,125
	60	2.01	174,180
	70	2.03	170,430
	80	2.01	179,233
	90	1.00	19,008
	100	1.99	199,321
200	50	2.01	180,821
	60	2.01	179,535
	70	2.10	258,302
	80	2.19	270,897
	90	2.04	172,299
	100	1.99	192,171
400	50	2.00	195,539
	60	2.05	174,848
	70	2.02	199,719
	80	2.00	195,231
	90	2.00	194,572
	100	2.00	186,047
600	50	2.05	157,244
	60	2.04	171,805
	70	2.02	178,812
	80	2.01	184,185

Max iteration	Max columns in X or Y	FoS	Volume (m^3)
	90	2.00	200,779
	100	2.00	195,602
800	50	0.82	38,239
	60	2.01	192,786
	70	2.01	204,299
	80	2.01	196,060
	90	2.01	187,613
	100	2.01	187,709
1000	50	2.02	185,006
	60	2.04	173,577
	70	2.03	175,154
	80	2.01	187,845
	90	2.01	189,787
	100	2.01	177,988

Table 2.
The influence of max columns in X or Y and max iteration in determining safety factor.

Figure 13.
3D slope geometry model.

Input parameter	Limonite	Saprolite	Bedrock	Unit
Unit Weight	17.19	15.77	29	γ_n (kN/m^3)
Cohesion	41.14	38.06	3270	C (KPa)
Internal Friction Angle	15.67	12.58	41.68	ϕ (°)
Modulus Young	20,000	20,000	150,000	ε (Kpa)
Poisson Ratio	0.30	0.40	0.20	ν

Table 3.
Material properties.

elements and nodes used will also be high so that the computation time can run for days depending on the number of nodes and types on mesh used. The computation time is closely related to the maximum number of iterations in the calculation of the

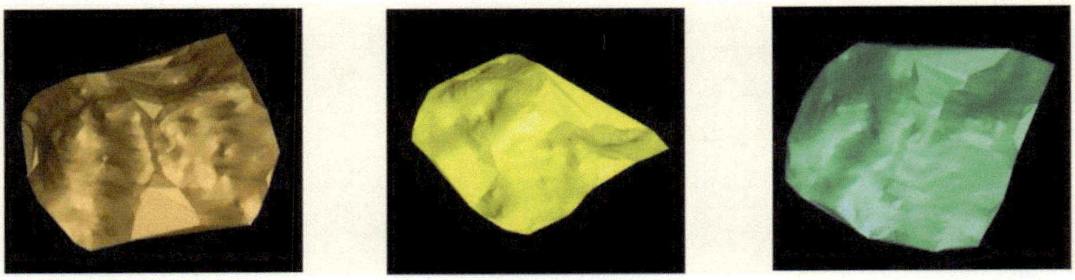

Figure 14.
3D geological floor model.

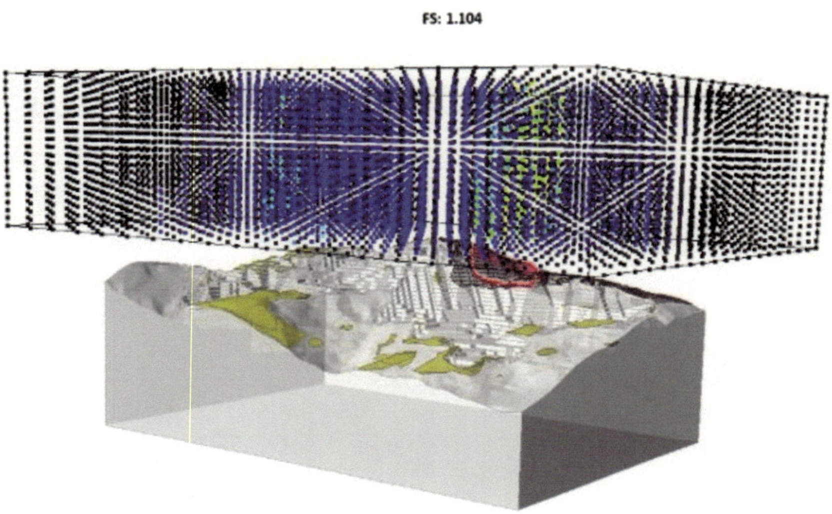

Figure 15.
Result 3D analysis using grid search.

Figure 16.
Result 3D analysis using cuckoo search.

force error or load imbalance (solid tolerance) in determining the convergence of model, the higher the maximum number of iterations, the calculation in determining the convergence level in the analysis result model will be more accurate, but unfortunately it will take a long time in the analysis.

Slope stability analysis using the finite element method takes into account the stress–strain analysis that works on each element of the model and focuses more on the analysis of the value of the deformation that occurs rather than the level of slope stability [2]. The advantages of analysis using the finite element method when compared to the limit equilibrium method are as follows:

- No need to assume the position of the slip surface, failure occur in zones where the shear strength of the material cannot maintain stability, due to the shear stress that works due to gravity.
- It does not require the concept of slice or columns, so it does not require an approach of forces acting on global equilibrium.
- Analyze the stress–strain so that it can see the deformation and the effective stress.
- The finite element method can monitor the movement of rock masses towards failure.

2.3.1 Strength reduction factor for 3D slope stability

The value of the safety factor (SF) in the finite element method is defined as the ratio of the actual material shear strength to the shear strength of the material when the model failure. This concept is similar to the limit equilibrium method, where the shear strength ratio of a material to its driving force [1]. The strength of the material at failure can be written in the following equation:

$$c_f = \frac{c}{SRF} \tag{43}$$

$$\phi_f = \arctan \frac{\tan \phi}{SRF} \tag{44}$$

With strength reduction factor (SRF) is the value of the factor for the decrease in the strength of the material, c_f is the cohesion value when the model failure and ϕ_f is the value of the internal friction angle when the model failure. In determining the strength of the material at failure, the technique of shear strength reduction (SSR) is used, where the actual material strength parameter is decreased step by step until the model become failure condition (non-convergent) [14]. The value of the material strength reduction factor for the Mohr-Coulomb criteria can be determined as follows:

$$\frac{\tau}{SRF} = \frac{c}{SRF} + \frac{\tan \phi}{SRF} \tag{45}$$

$$\frac{\tau}{SRF} = c_f + \tan\ \phi_f \tag{46}$$

The systematic stage in the model analysis to determine the critical value of the material shear strength reduction factor (Critical SRF) is as follows:

- The first stage is to prepare the model for analysis using the finite element method, determine characteristics of strength and deformation properties of the material.
- The second stage is to increase the value of the material strength reduction factor (SRF) and calculate the material parameters using the Mohr-Coulomb criteria, then input the new material properties data into the model and recalculate and record the maximum total deformation value.
- Repeat the second stage using a systematic increase in the value of the material strength reduction factor, until the model becomes a non-convergent condition (failure). The critical SRF is determined at the highest srf value in the model to achieve the convergent condition.

3D model analysis is an extension of 2D analysis, but in modeling complex geometries 2D analysis cannot simulate them properly. The 2D analysis assumes that the slope width is infinitely wide [11]. However, in many cases the 2D analysis is considered more conservative because it results in a lower safety factor value compared to the 3-dimensional analysis. The weakness of the analysis which is carried out in 3 dimensions is that it takes a lot of time and money, which makes practitioners not want to switch. The results of the 3D analysis can be conservative if they are validated with a 2D analysis in the most critical areas [2]. For conservative results, the 2D and 3D analyzes should not be significantly different, however in many cases the 3D analysis provides a higher safety factor value. The advantage given if the analysis is carried out in 3D is that it can represent the actual slope conditions in real terms and can determine the critical position. The value of the safety factor in the finite element method analysis with the 3-dimensional model is determined by the use of the material shear strength reduction factor (SRF) technique. in order to obtain the true safety factor, the srf is gradually increased until the model become failure (non-convergent). When this critical value is found, the safety factor of the slope model is equal to the reduction in material strength (SF) ≈ (SRF).

The determination of the critical value of srf is determined by increasing and decreasing the SRF value step by step until the highest SRF value is obtained in the model to be able to achieve the convergence criteria, see **Table 4**, the slope model enters a non-convergent (failure) at srf 1.05, so that the critical value of srf is 1.04 and highlighted in bold color.

2.3.2 Mesh type

Finite element analysis method employs the utilization of elements and nodes to perform stress–strain analysis acting on each element. The process of establishing

Step	SRF	Solid Tolerance	Convergence
1	1	0.0005	Yes
2	1.3	0.2813	No
3	1.14	0.0691	No
4	1.06	0.0152	No
5	1.02	0.0008	Yes
6	**1.04**	**0.0009**	**Yes**
7	1.05	0.0078	No

Table 4.
Critical SRF determination.

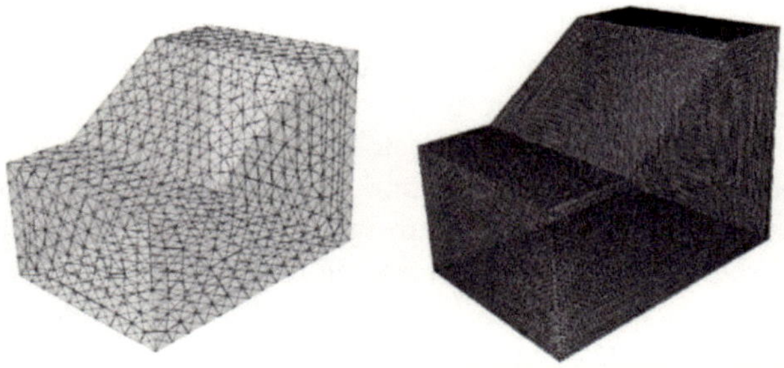

Figure 17.
4-node graded and uniform.

these elements is called meshing. Mesh type will affect the analysis result. Greater number of elements will lead to a more accurate analysis result. However, this will result a more time-consuming computation. There are various mesh types in 3-dimensional analysis, there are 4-nodes and 10-nodes with uniform and graded shape available to this method. The example of elements and nodes utilization in FEM are given in **Figure 17**.

To recapitulate the results of the analysis can be seen in **Figure 18**. From this figure provides information that there is an effect of using the mesh type on the srf value of the analysis results obtained, this is because the number of elements and nodes used is also different.

2.3.3 Maximum iteration optimization

In model analysis using the finite element method, the determination of the convergence criteria is limited by the analysis calculation (maximum iteration), the more iterations allowed, the more accurate the analysis results will be, but it also needs to be considered in terms of the slope model being analyzed, in the use of the maximum number. Optimal iteration related to the level of efficiency in the analysis so as not to waste too long.

At first, the model will be analyzed the stresses that work on each element due to the applied load, it will get the principal major stress and minor for each element,

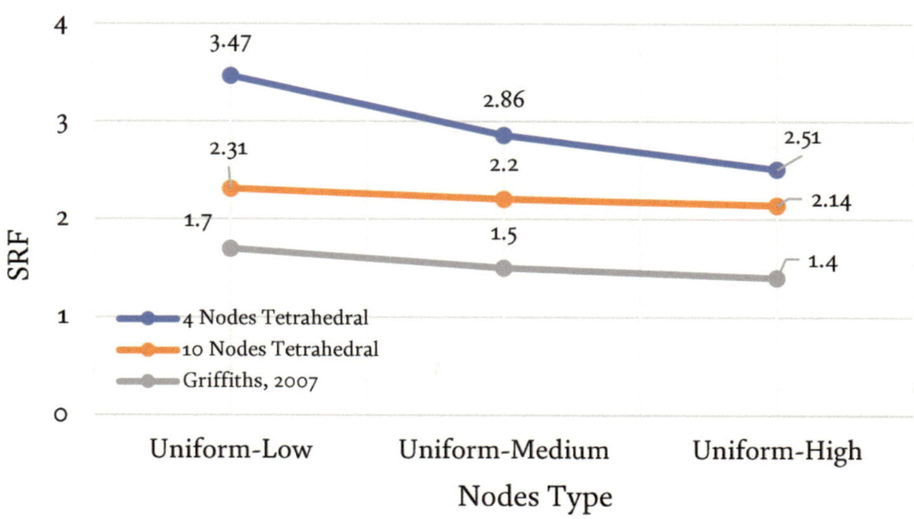

Figure 18.
SRF VS mesh type.

then enter the material properties for each material and start entering the material strength reduction factor (SRF) value for each element, at this stage the material properties value can be increased or decreased depending on the error value (solid tolerance) obtained. Furthermore, an Elasto-plastic analysis was performed using the Mohr-Coulomb failure criterion to obtain a new error value (solid tolerance). If the error value is still above the maximum value within the allowable iteration calculation limit, the SRF value will be lowered until the error value is below the maximum limit. The recapitulate the analysis results can be seen in **Table 5.**

The analysis shows that the higher the maximum number of iterations, the SRF value also increases **Table 4**. As the SRF value increases, in the maximum number of iterations of 400 the SRF value constant at a value of 1.04, so this provides information that at a maximum of 400 iterations is the maximum optimal number of iterations. If seen from the effect of the number of iterations on the total displacement, the results will fluctuate, this is closely related to the srf value because the higher the srf value, the total displacement will also increase, but at 1.04 the srf value does not change/is consistent but the total displacement fluctuates because it is solid tolerance, the resulting solid tolerance also varies due to the use of different maximum iterations, this is because the slope conditions remain non-convergent (energy equilibrium is not achieved) above srf 1.04 in the number of iterations that have been set, however, solid tolerance will get closer to the maximum value, while for the computation time it is very clear that the higher the maximum number of iterations, the computation time will also increase, because it will take more time to search for convergence with the maximum iteration limit given, although the results will remain the same, that non-converging above 1.04 and the last converging value is 1.04.

2.3.4 Analysis 3D finite element method of open pit mine

Analysis using the finite element method uses the same data as the analysis carried out with boundary equilibrium, namely rock characteristics, 3D slope

Max iteration	SRF	Max displacement (m)	Computing time (s)
10,000	1.04	8.60	110,459
5000	1.04	8.74	57,990
2500	1.04	8.48	26,214
1000	1.04	24.29	14,622
500	1.04	14.34	6021
450	1.04	12.90	5422
400	1.04	12.72	4788
350	1.03	5.25	3650
300	1.03	5.42	3341
250	1.03	6.51	3159
200	1.01	1.70	2300
150	1.01	1.70	1969
100	1.00	1.67	1704
50	0.98	1.20	1541
25	0.68	1.09	1501

Table 5.
Influence of variable max iteration.

geometry and 3D geological models. Young and poisson ratio are input parameters in this analysis. If the boundary equilibrium analysis requires the position and shape of the slip plane, the analysis does not require these assumptions but uses the elements and nodes that are attached to the model. For an example of a 3D mesh model, see **Figure 19**. For this example, case a 4-node element type with a graded gradient is used.

Figure 19.
3D meshing model.

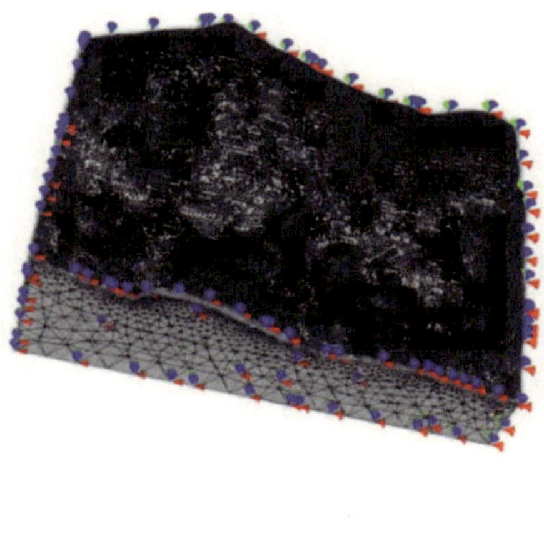

Figure 20.
3D restraint model.

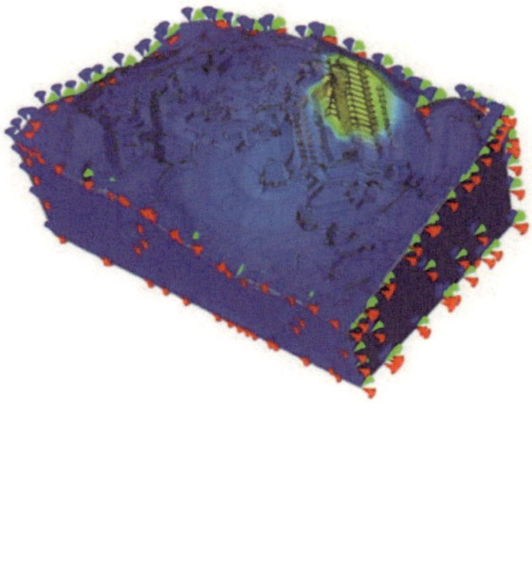

Figure 21.
3D FEM analysis result.

After the model is successfully meshed, it is necessary to determine the limit of the resistance that works in the field of analysis to determine the internal force that works on the restraint installation which plays a role in determining the deformation limit of the model. For an example of the restraint model can be seen in **Figure 20**.

Analysis with the finite element method can see information about stress acting on the model and the total displacement (**Figure 21**).

3. Conclusions

Slope stability using LEM shows that the cuckoo algorithm is reliable in obtaining position and shape of slip surface. In finite element analysis method, the optimum iteration number needs to be considered to improve analysis efficiency and preserving accuracy.

Acknowledgements

The authors would like to thank PT X who has facilitated the collection of field data, as well as the Mining Computation Laboratory of Mining Engineering Department FTKE of University Trisakti which has facilitated the author in using the software used in modeling research data.

Author details

Masagus Ahmad Azizi[1]*, Irfan Marwanza[1], Muhammad Kemal Ghifari[1] and Afiat Anugrahadi[2]

1 Mining Engineering Department FTKE Universitas Trisakti, Jakarta, Indonesia

2 Geology Engineering Department FTKE Universitas Trisakti, Jakarta, Indonesia

*Address all correspondence to: masagus.azizi@trisakti.ac.id

References

[1] D. V. Griffiths and R. M. Marquez: Three-Dimensional Slope Stability Analysis by Elasto-Plastic Finite Element. DOI: 10.1680/geot.2007.57.6.537

[2] Duncan, J. M: State of the Art: Limit Equilibrium and Finite-Element Analysis of Slopes," Journal of Geotechnical Engineering Division, ASCE, Vol. 122, No. GT-7, pp. 577-596

[3] A nagnosti, P: Three-dimensional stability of fill dams. Proceedings of the 7 th International Conference on foil Mechanics and Foundation Engineering, Mexico, Vol. 2, pp. 275-280

[4] Baligh, M. M. and A zzouz, A. S: End effects on stability of cohesive slopes, "Journal of the Geotechnical Engineering Division, AS- CE, GT 11, pp. 1105-1117

[5] Hovland, H.J: Three-dimensional slope stability analysis method," Journal of the Geotechnical Engineering D iv., AS CE, Vol. 103, GT 9, pp. 971-986

[6] Hungr O, Salgado FM, Byrne PM. Evaluation of Three-Dimensional Method of Slope Stability Analysis. Canadian Geotechnical Journal. 1989;**26**: 679-686

[7] Stark, T.D., and H. T. Eid: Performance of Three-Dimensional Slope Stability Methods, Journal of Geotechnical Engineering Division, ASCE, Vol.124, No.GT-11, November. 1049-1060

[8] Y. M. Cheng and C. J. Yip: Three-Dimensional Asymmetrical Slope Stability Analysis Extension of Bishop's, Janbu's, and Morgenstern-Price's. DOI: 10.1061/(ASCE)1090-0241(2007)133:12(1544)

[9] Lee W. Abramson, Thomas S. Lee, Sunil Sharma, Glenn M. Boyce. Handbook of Slope Stability and Stabilization Methods. 2nd ed. ISBN: 0-471-38493-3

[10] Azizi. M.A, Kramadibrata. S, Wattimena. R.K, Indra. D.S, Ardiansyah: Analisis Risiko Kestabilan Lereng Tambang Terbuka (Studi Kasus Tambang Mineral X) Geotechnical Superintendent PT Newmont Nusa Tenggara. In: Proceedings of Workshop Symposium National Geomechanics-1; June 2012; Yogyakarta

[11] Azizi. M.A, Marwanza I, Anugrhadi A, Fradiba A.A, Amelia N: The Influence of Number of Grid Point and Radius Increments in Determining Safety Factor and Estimated Sliding Volume on Three-Dimensional Slope Stability Analysis. DOI: 10.1088/1757-899X/478/1/012041

[12] John Read and Peter Stacey. Handbook of Guidelines for Open Pit Slope Design. ISBN: 9780415874410

[13] Yang X-S and Deb S. Cuckoo Search via Lévy Flights. IEEE Publ 2009:210–4. DOI: 10.1109/NABIC.2009.5393690

[14] R. Hammah, T. Yacoub, B. Corkum, F. Wibowo, and J. H. Curran: Analysis of Blocky Rock Slopes with Finite Element Shear Strength Reduction Analysis. DOI: 10.1201/NOE0415444019-c40

Chapter 5

Asphalt Fill Strengthening of Free Slip Surfaces of Shale Slopes in Asphaltite Open Quarry: Stability Analysis of Free Sliding Surface for Wet Shale Slopes in Avgamasya Asphaltite Open Quarry No 2. Site

Yildırım İsmail Tosun

Abstract

The stability analysis carried out by GEO5 software and uses free sliding analysis by wet and pore saturated weight charting provided the safety factor of 1.35. The safety precautions were followed by inclinometers and wire extensometer measurements. The other pore saturated asphalt bound shear box and unaxial test compression tests were resulted in the geotechnical and geoseismical data over sliding soil /shale inter surface quality and the characteristics of free rock falling risk and discontinuity distribution, sub crack density and distribution on stereo nets were determined. The research was firstly followed the perched water levels on geoseismical data over causing water burst or explosion of highly free mud and landslides. The hazardous rock falls over saturated soil and uncohesive rock explosions. The proposed study was secondly as strengthening methods such as asphalt mixing as precautious on shear stabilization and other wire mesh barriers anchored. The free sliding cracks was filled by asphalt and compressed for stabilization strengthening known as the characteristics of avoiding shear falls in the future. The unconditional expectations related to this study was also defined for this region such as the influence of the ground water, rock cracks and slope design, explosion exchange dynamics leading to landslide.

Keywords: mining pit, Şırnak asphaltite, free slide, landslide, mining, geoseismical analysis, stabilization, shale stability, slope stability

1. Introduction

Open pit mine slope stability and the rock fall risk assessment were studied in high steep excavation site in order to avoid landslides or rock falls occurred several times in Avgamasya Asphaltite mining sites, Şırnak. The high steep slopes were reaching over 120 m high with partly 60–65 degree shale/soil slopes developing major free landslide hazard in harsh climate conditions in recent years. The coal seam was so vertical diving at the working area of miners, made compulsory to pre-search and take precautions by a hot melted asphalt/fly ash mixture filling through tension cracks over rock falling and free sliding cavities in the hazardous mining operation area in the open pit asphaltite mining. The critical issues and precautions for free failure in Avagamasya Open Pit Site No 2 were mentioned as below:

- Basic instability parameters searched by many researchers mentioned above anisotropy as often caused sudden failure by hazardous fractures and cracks in excavation and mining sites [1–10].

- Tensile cracks and shear loads that occurred in the free slip surfaces on heterogeneous breccias formations in Şırnak geology [11, 12] and cracks and land of asphaltite mining investigated by standards [13–16].

- Hazard risk can be examined as water income and levels in the mine [17, 18]. In the pit slopes, these values of pore pressure u are synchronized the stress and the ground water income patterns should be extracted [19, 20].

- During the works, asphalt filling stabilization of cracks on the safety of the slopes [21–24] and steps are formed in the work stages, the safety of the truck transportation road and the safety of the excavation area were ensured by anchorage reinforcement [25–29].

- A steeper safe stepping of the slopes and a suitable minimum fly ash addition pouring and crack area of asphalt filling design has been developed lower horizontal adhesive stabilization and less cracks [30–36]. High compacting vibration, drainage and low excavation capacity were also affected excavation time, free slope, crack yields and steep displacements following discontinuous failures [37, 38].

Ideally, extensometer patterns and wire cables in the stereo net were explored awareness of slide by examining the fractures in the Avgamasya asphaltite mining area. Tele monitoring of boreholes and level mirrors was made daily. In addition, acoustic sound noise analysis was also carried out. Since drilling inspection holes and daily wire follow-up was costly, taking measurements with tension wire extensometers in bevel chucks showed more reliable horizontal deformation values. Accordingly, 45 and 50 degree angles of Şırnak asphaltite quarries were found to be reliable [28, 32]. Two different design models have been developed in order to obtain these stability values. According to these models, it is thought that the slopes may be exposed to planar and wedge type shifts and frees sliding over the slip planes depending on the fracture bedding and density (**Figures 1** and **2**).

The observation of different asphaltite qualities in the field and the diversity of production provide the identification of qualified coal seams needed. Depending on the strength and hardness properties of coal, it is more difficult to determine the chemical structure and strength of asphaltites and side rocks containing heterogeneous structures compared to many other types of coal. The strength and failure type of the country rock differ. The components that make up the coal differ depending on the distribution, orientation, amount and strength and hardness values of these components in the asphaltite sample. This situation is determined relatively within a homogeneous asphaltite matrix. More qualified asphaltite shows softer mechanical strength. Hard veins are also very important in the country rocks. Asphaltite quality and development can be determined underground by seismic reflection and resistivity measurements, depending on the density **Figures 3** and **4**.

Asphaltite and shale, marly shale, marly limestone and various marly formations that are the subject of this study can also be revealed. The Siirt Formation, which is

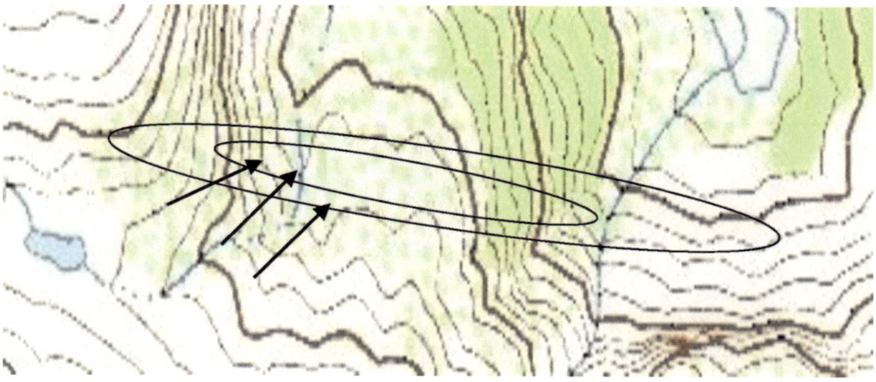

Figure 1.
Avgamasya location of Asphaltite quarry No 2 in Şırnak.

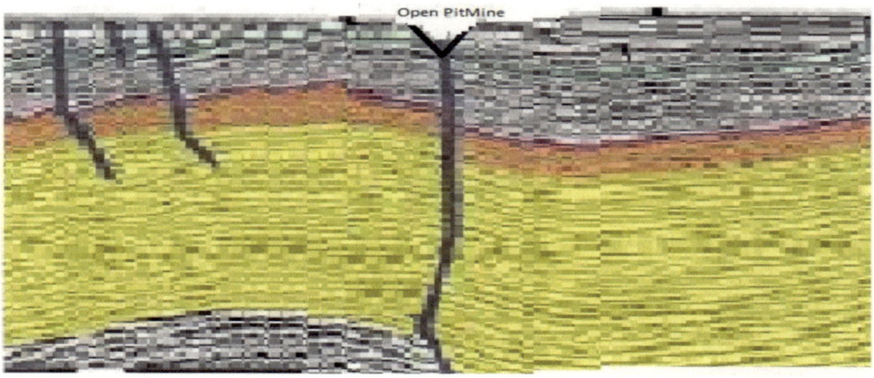

Figure 2.
Seismical formation layers of Asphaltite quarry No 2 in Şırnak.

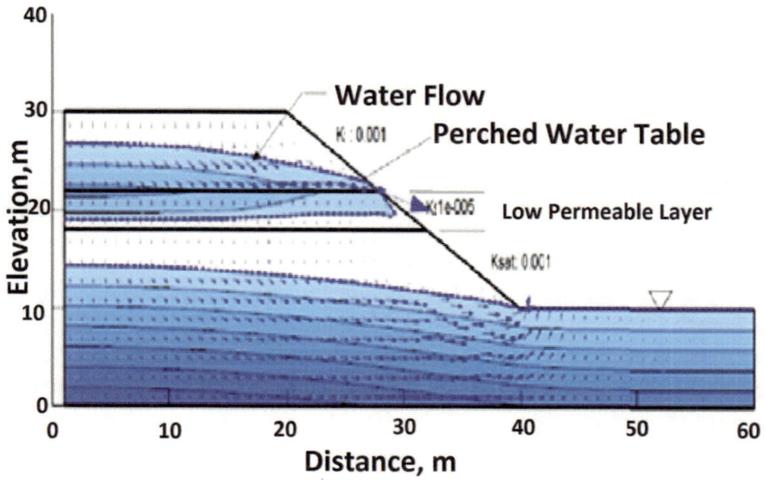

Figure 3.
S1 and S2 slope free slip surface perched water table through free slip surface.

outcropping in the west–east directions of Şırnak province, is important in bedding and forms important rock units in asphaltite bedding in terms of rock mechanics. This unit generally consists of marls containing clay and shale and clays with asphaltite. Sulphate-rich water resources are also located in these units consisting of

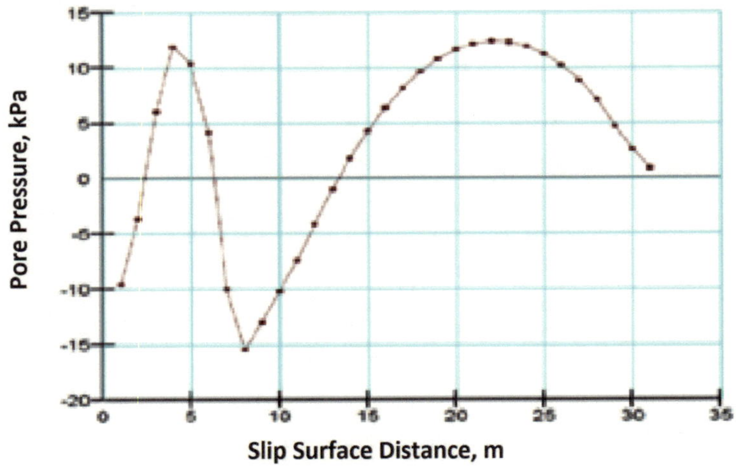

Figure 4.
S1 and S2 slope free slip surface slip surface water saturation.

claystone rocks. Due to its different mechanical strengths and densities, it helps to find underground bedding. In addition, it has been evaluated within the scope of this study since it creates various engineering data.

In operation, the drilling cores are subjected to mechanical tests on site and under laboratory conditions. The micro structural and mineralogical studies carried out. Compression tests were carried out on samples with relatively weak asphaltite and claystone levels and hard, higher strength asphaltite veins with diameters of 34 mm and 84 mm.

With the findings obtained, it was determined that asphaltites containing components with different strength and hardness properties show different fractures. It is aimed to determine the changes in different breaking stresses and strengths for each borehole and logs.

Possible bedding will be examined with seismic reflection method and exploration drillings will be opened at the location points displayed on the map below. Drilling exploration locations will be opened at 1000 m depth at the points shown on the map. Seismic reflection data describe possible asphaltite bedding as shown in the **Figure 2**.

1.1 Uniaxial compression strength tests

Uniaxial compression strength tests without environmental stress were carried out in the Şırnak University laboratory with a 2000 kN capacity hydraulic controlled test device. Axial and circumferential deformation gauges are placed at the levels of half of the sample lengths, as shown in the figure, so that their measurements are not affected by changes in the sample edges. The axial strain gauges were placed tightly on both sides of the specimens, mutually. Measuring range of axial deformation gauges on the sample is 50 mm for 84 mm diameter samples and 35 mm for 34 mm diameter samples as shown in **Figures 5** and **6**. The circumferential strain (rad) was calculated with the help of strain gauges connected to the chain wrapped around the sample. During the test (**Figure 7**), digital feedback was provided by circumferential strain and the control value used was set at 0.05 mm/min. The rock properties were defined.

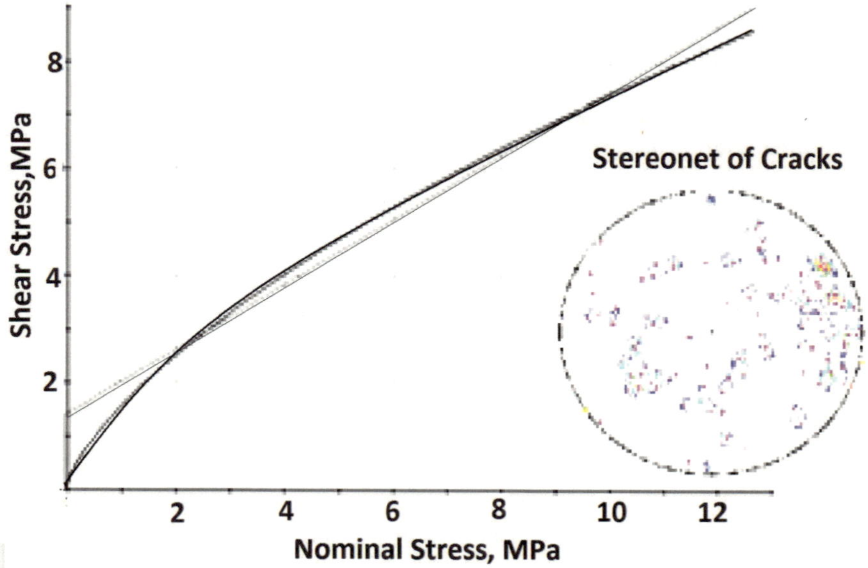

Figure 5.
Typical Mohr coulomb envelope of the Asphaltite mine shale rocks studied.

Figure 6.
Typical chlorite and calcite belts of the Asphaltite mine shale rocks studied.

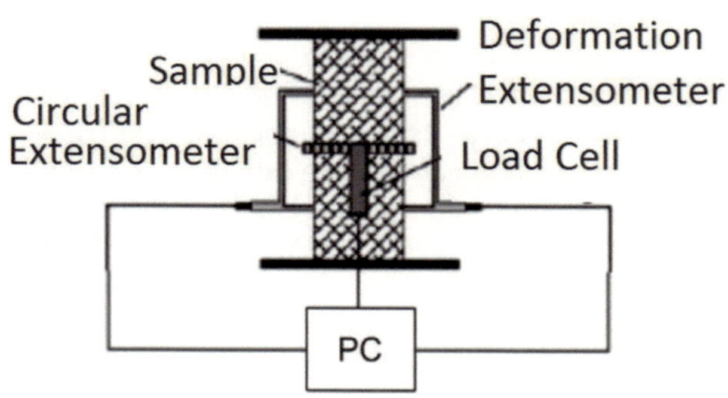

Figure 7.
Schematic view of the unaxial compression strength test setup.

2. Stability modeling and asphalt covering for free rock sliding control in Şırnak open pit coal mining

Water drained layered surfaces confirmed the geotechnical stability caused anisotrophical pore pressure developed and compression strength developed. The below equation showed the shear on fre slip surface with cohesive mater.

$$c_{\theta} = c_2 + (c_1 - c_2)\ \cos^2\theta \tag{1}$$

where the shear force and stress and nominal load and compression stress at angle was θ caused failure at weak layer shear in the slope.

Vertical unaxial strength and Horizontal strength of the specimens were changed.

GEO5 model weight slice chart construction carried out as given below serial equation sum:

$$F = \sum_{0}^{\mathrm{i}} \mathrm{N}_i\mathrm{F}_\mathrm{i} = \mathrm{N}_i \frac{C_\mathrm{i}}{\gamma H} \cos\beta^i \tag{2}$$

N_i slice load kN $\frac{C_1}{C_i}$, β slope angle, The slip surface stability analyzed as load chart sum. Safety factors over 1,35 and 1,5 was confirming the stability.

Regarding the crack distribution and density orientation and intersection with water perched tables shown (**Figure 5**) as shear risk factor $\mathrm{R}_{c.}$

10 m length slice at i discontinuity at angle of crack and density of % heteregenous distriburtion on slip surface as $\frac{dy}{dx}$ was calculated.

$$\mathrm{R}_c = \sum_{0}^{\mathrm{i}} \mathrm{R}_i\mathrm{F}_\mathrm{i} \tan\theta = \int_a^b e^{-ti\theta} dy^i \tag{3}$$

$$\frac{dy}{dx} = e^{-ti\theta} dy \tag{4}$$

The studied stages were as below:

Slope Stability Chart modeling was managed as shown in **Figure 3**.

The Stability mechanism and control by asphalt crack filling and cable net pillar construction was avoiding pore pressure for each slice as given Eq. (5)

$$\frac{dy}{dx} = u = \sum_{0}^{\mathrm{i}} \mathrm{R}_i\mathrm{F}_\mathrm{i} / \tan a \left(1 - e^{-tRi/\mu}\right)^i \tag{5}$$

u deformation intrinsic friction resistance, F weight slice, a shear fracture inclination angle t time, μ crack mud viscosity i weight slice.

The safety factor in free sliding has been investigated by following the stress cracks by transforming the slope deformations based on the internal friction angle patterns. Fracture agglomeration and cohesion-free free fall displacements can be observed above 50 mm. In order to increase the viscosity in the cracks, the waste liquid polymer materials were poured into these gaps to provide stability. The joint density over slip surface for each slice was calculated by the Equations sequentially as below:

$$\mathrm{Ji} = \sum_{0}^{\mathrm{i}} \mathrm{N}_i\mathrm{F}_\mathrm{i} \tan a_i \tag{6}$$

$$\mathrm{Ri} = \sum_{0}^{\mathrm{i}} \mathrm{S}_i \mathrm{W}_\mathrm{i} \cos a_i = \tag{7}$$

$$\mathrm{p_u} = \frac{\Upsilon'}{\Upsilon} \mathrm{H}_i \tag{8}$$

$$\mathrm{Fi_u} = \sum_{0}^{\mathrm{i}} \mathrm{W_i} \sin a_i - \mathrm{S}_u = \sum_{0}^{\mathrm{i}} \mathrm{W_i} - \mathrm{p_u} \tag{9}$$

$$\mathrm{S_{iu}} = \sum_{0}^{\mathrm{i}} \mathrm{c}_u{}'l \ sec\, a_i + \mathrm{Fi_u} \frac{\Upsilon'}{\Upsilon} \mathrm{H}_i \tan^2 \phi' \leq 1,25 \tag{10}$$

$$\sigma'_\theta = \sigma - u_a + \chi(u_a - u_{wi}) \tag{11}$$

$$\tau_{\theta i} = c'_i + (\sigma - u_a + \chi(u_a - u_{wi})) \tan\phi' \tag{12}$$

$$\mathrm{S_{iu}} = \sum_{0}^{\mathrm{i}} \mathrm{c}_u{}'l \ sec\, a_i + (\sigma - u_a + \chi(u_a - u_{wi})) \tan^2 \phi' \leq 1,25 \tag{13}$$

safety water saturated rock parameters regarding pore water content

$$\mathrm{M} = \cos a \left[1 + \frac{(\tan a \tan \phi')}{F} \right] \tag{14}$$

$$\mathrm{R_u} = \frac{\Upsilon_w{}' h_i}{\Upsilon d} \tag{15}$$

Safety factor was calculated by perched water table and water saturation

$$\mathrm{S_{Rock}} = \mathrm{S_u} \sum_{1}^{\mathrm{i}} \{ [\mathrm{c'b} + (\mathrm{W_i} - ub) \tan \phi'] / \cos aM \} / \sum\nolimits_{i=1}^{n} \mathrm{W}_{iu} \ sin\, a_i \tag{16}$$

$$\mathrm{S_{Rock}} = \frac{2c}{\Upsilon H} PiRi(Q_i' \cos \phi - R(P+S)_i / (Q_i' + RS_i \cot \phi)) \leq 1,25 \tag{17}$$

$$\mathrm{P_i} = \left\{ 1 - \frac{z}{H} \right\} \ \mathrm{cosec}\, \phi \tag{18}$$

$$\mathrm{Q_i} = \left\{ \left[1 - \frac{z}{H} \right]^2 \cot \phi - \cot \phi \right\} sin\phi \tag{19}$$

$$\mathrm{R} = \left(\frac{\Upsilon_w}{\Upsilon} \right) \left(\frac{z_w}{z} \right) \left\{ \frac{z}{H} \right\} \tag{20}$$

$$\mathrm{S_i} = \left(\frac{z_w}{z} \right) \left\{ \frac{z}{H} \right\} \ \sin \phi \tag{21}$$

$$\mathrm{D_i} = \left\{ \left[1 - \frac{z}{H} \right]^2 \cos \phi \right\} \ \cot \phi \ (\cot \phi \ \tan \phi - 1) \tag{22}$$

2.1 Rock texture, mineralogy and petrographic characteristics

As seen in **Figure 8**, the samples subjected to the tests generally present a heterogeneous rock texture consisting of claystone levels and anhydrite veins and layers of different frequencies. The thickness and elongation of the calcite veins in the claystone levels range from mm to cm. Its mineralogical composition is

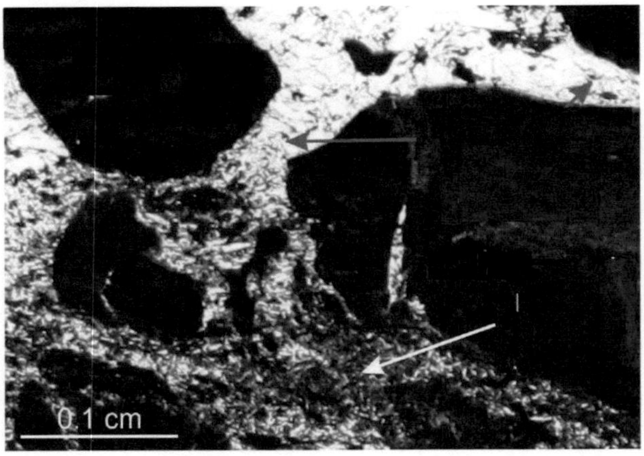

Figure 8.
Different chlorite shale matrices according calcite to shale content.

generally composed of anhydrite and clay minerals. Typically, the clay content and anhydrite content are inversely proportional to the samples. In other words, samples with low clay content generally contain higher density calcite veins.

Changes in the microstructure and mineralogy of the clay matrix, which appeared homogenous macroscopically, were determined in the microscopic examination of the samples subjected to experiments. As seen in **Figure 5**, the brown, dark brown and black regions usually contain clay matrix and rarely scattered calcite lumps in it. The areas seen in more gray tones form fine-grained anhydrides dispersed in fine-grained clays and form calcite-rich clay matrix. The anhydrite content and distribution in the clay matrix varies in sections. Besides, different types of matrix can be found in the same example.

As a result of the experiments conducted under environmental stress conditions, it was determined that the highest strength values σ varied between 9 and 81 MPa depending on the mineralogical composition of the samples. Failures occurred along the clay matrix in samples with low uniaxial compressive strength values. The samples with high values σ are the samples containing more than 84% marl. The general images of these examples show a homogeneous structure. The uniaxial compressive strength values σ obtained were accepted as extreme values σ indicating the strength properties of marl veins and clay matrix. In cases where the clay content is more than 10%, the uniaxial compression strength decreases very little. The opposite situation develops when the clay content is less than 7–10%. In this case, uniaxial compressive strength increases significantly due to the relatively increasing marl veins.

After these results, the relations between mass distribution of clays and σc under uniaxial conditions are evaluated in **Figure 9**. The standard test results similar to the σ values were obtained in uniaxial compression strength were encountered. When the clay content is more than 7–10%, the σc values decrease with an almost constant orientation.

Thus, in conditions where the clay content is less than 7–10%, σc values τ show a significant increase with the decrease of the clay percentage. In view of these results, it is thought that the initial fractures were controlled by the mechanical properties of the claystone levels, as the claystone levels were weaker than the low calcite belt levels.

The presence of chlorite belts is shown in claystone matrices of different type belts and different anisotropic strength properties together in the same sample. In the slip area of Avgamasya Open Pit quarry No 2. shale face the sliding shale formations is shown in **Figures 10** and **11**.

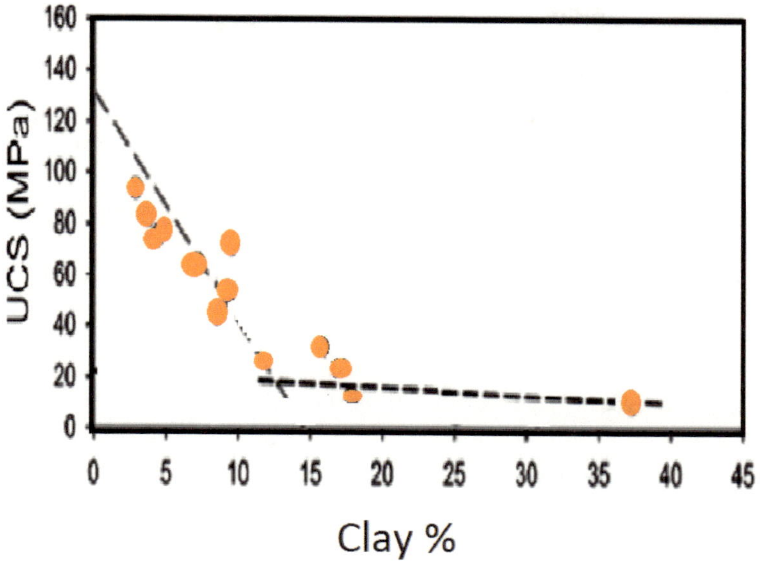

Figure 9.
Change of σ uniaxial compressive strength value according to clay content.

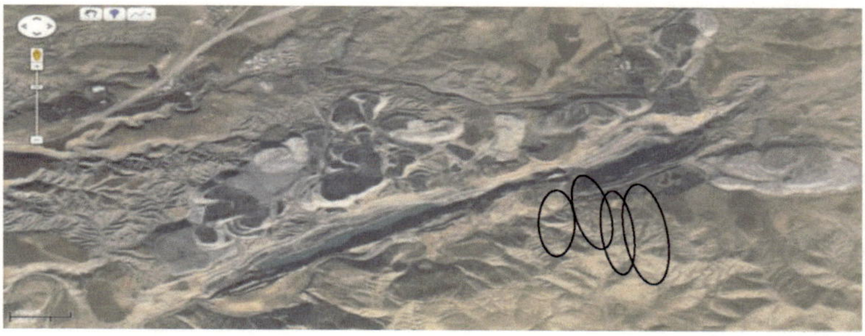

Figure 10.
Free rock slide and falling site in Şırnak Asphaltite open quarry No 2, satellite view 1/18000.

Figure 11.
Free rock slide and falling site in Şırnak Asphaltite open quarry No 2, free slide over excavation area, 1/1000.

3. Geotechnical studies

Stack weight in the slip area to determine the geotechnical stability in American Standards (ASTM and GEO5) based. In the area where the strengthening cabling of the masses and the presented soil rock interface of the phase content is given in **Tables 1** and 2.

Slip face of shale and limestone rock pillar parameters are given in **Table 3** below.

Free slide rock stability to assess the risk of slipping GEO5 program used and was advantageous. Rock stability GEO5 program provided safety analysis at 1.35 safety factor. Stability analysis of weight slices as wide 3 m blocks as possible cut into slices. On Slope S1 free slip surface formed like 2m mesh using GEO 5 FEM program by groundwater data section submerged discharge gave the critical red

Sample No	Asphalt Ash compost fill	Asphalt+Fine Shale compost fill	S1	S2	S3	S4
height(m)	800	850	925	921	933	927
Wopt,%			15.90	13,70	10,80	11,40
c′(kpa)	52	88	0,52	0,59	0,63	0,55
φ′	24,2	22,5	32,50	22,50	21,00	20,00
L1(%)	11.8 Mpa σ	9.6 Mpa σ	26	15	28	17
P1(%)	42 RQD	40 RQD	19	11	18	22
Ip (%)	46 RMR	44 RMR	10	9	8	12
γs g/cm^3	2,70	2,70	2,40	2,50	2,40	2,30
Soil	zayıf	zayıf	SP	SP	SP	SP
$\gamma_{doğal}$ g/cm^3	1,94	2,14	1,82	1,76	1,90	1,70
γ_{kuru} g/cm^3 Kum ve çakıllı	1,94	2,14	1,65	1,6	1,78	1.60
γ_{doygun} g/cm^3	2,0	2,23	2,02	1,84	2,0	1,8

Table 1.
The samples taken from the slopes of the masses on the results obtained from the geotechnical testing.

Örnek no	S11	S21	S31	S41
γ_k max g/cm3	1,68	1,93	2,05	1,90
wopt %	15,9	19,0	12,3	13,0
Permeability (k) (cm/s)	5,63*10^{-4}	6*10^{-4}	3,0*10^{-4}	5,62*10^{-4}

Table 2.
Permiability of the samples taken from the slopes.

Specification of rock fill	Asphalt Ash compost fill	Asphalt+Fine Shale compost fill
Natural unit weight, γ_n (kN/m^3)	16	18
Saturated unit weight, γ_d (kN/m^3)	16	18
Cohesion, c (kN/m^2)	52	88
Intern. Friction Angle, ϕ (°)	30	34

Table 3.
Physical and mechanical properties of asphalt composite fill.

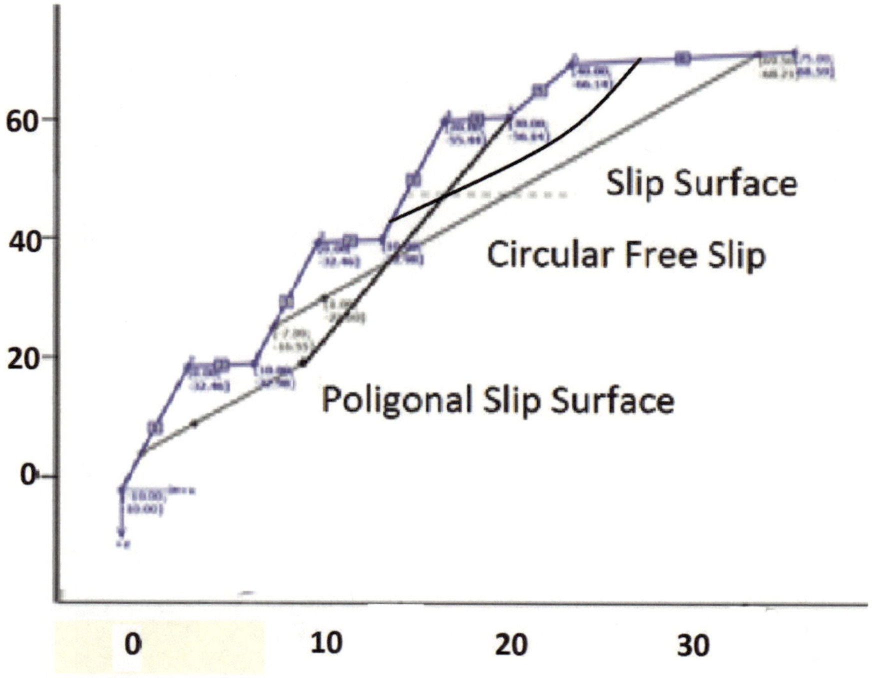

Figure 12.
S1 cross-section of rock slope of the study area, circular free rock sliding surfaces, sensitivity analysis rock stability using GEO5 program.

level slip surface effect and poor stability factor was practiced by lower mesh cross section [19, 20]. Using GEO5 on slip slice chart is advantageous at different slip surface pore pressure in program as in **Figure 12**, depending on the rock surface or planar wedge type drift is not formed in particular from 30 to 40 m length was determined.

Safety factor was in the ASTM standard as based GEO 5 rock stability safety factor of over 1.35. Using the parameters determined by model chart analysis of GEO 5, the appropriate support system for the long-term stability of the slopes was determined. Accordingly, it was decided to make asphalt cohesive support with fly ash material in front of the slope. Subsequently, 2 m long rock bolts were placed to provide short-term slope stability. After this stage, it was understood that a compost of asphalt-cover structure should be made on the crack section. The slope was required to ensure the long-term stability of the Asphaltite open pit quarry.

4. Slope analysis of S1 and S2 shale soil/rock face slopes

The stack S2 hillside after rains made the small size of the movements that have been identified in field studies. No. S2 to develop pile slope is covered with talus and 10 m mesh rock stability GEO5 programs were created problem due to heterogeneous structure and complexity with the program (**Figure 3A**). S2 The top of the stack and the maximum height difference between the heel point 45 m, the slope of the maximum height of 50 m, the slope of the surface tilt angle is 40°.

Regarding cohesive resistive parameters obtained from the asphalt composed rock formations made in c '= 1.9 kpa, φ '= 22°, $\gamma_{nat.}$ = 1.97 = 2.27 g/cm^3 g/cm^3, and are used to γ_{dry}. According to γ_{dry} and γ_{nat} calculated separately on the potential slip surface deformation iso values seen in **Figure 13**.

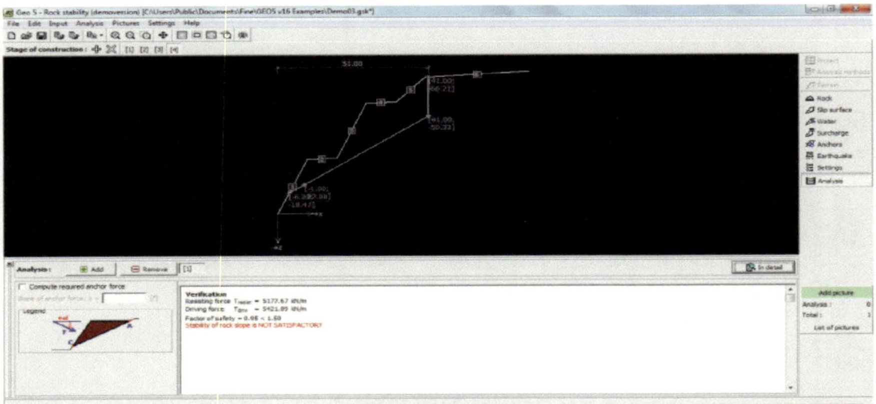

Figure 13.
S1 section of the study area slopes 10 m slice topology, b. Deformation rock stabilty stability analysis GEO5 programs, cut red 30 mm unstable displacement.

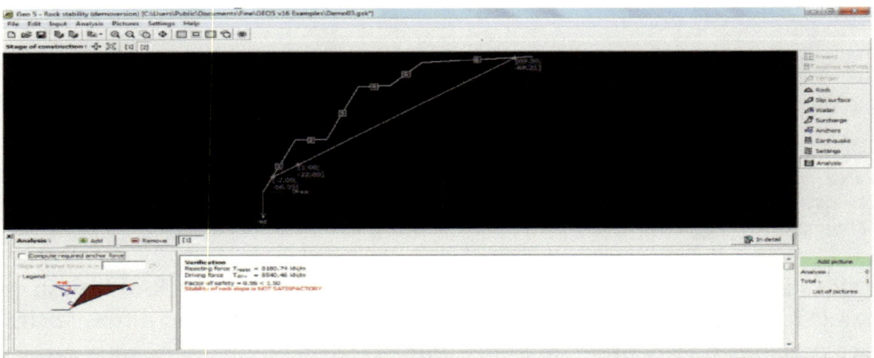

Figure 14.
S2 section of the study area slopes 10 m slice topology, b. deformation rock stabilty analysis GEO5 programs, cut red 30 mm unstable displacement.

S2 after water perched table on the slope of a deep instability, high shear force occurred. The instability was observed. Similarly the slope S3 indicates instability and displacement reached 30 m depth slip circular surface in Bishop Theory. (**Figure 12**)

In addition, 1 m wide and 10 m high pillar construction by cable cover by a safety factor values were above 1.5 (**Figure 14**)

When using a pillar to reduce instabilities under perched water tables in the slip deformation displacement was shown in **Figure 8** and displacement m below the maximum possible shift of the substrate reached 9 m depth.

The rock and filled asphalt waste fly ash compost shear stability ranged from 10 to 13 kPa with 610 kPa reaching a possible shift in the base.

5. Results and discussions

Therefore, the existing design was updated in this way and the safety coefficient of the slopes was obtained over 1.5 when the asphalt cohesive support structure was filled with 2 m filling as seen in **Figure 4**. As a result, the long-term security conditions of the shale slopes could only be achieved with these suggested measurements. and pillar cabled support. Cracks occurring in the S1 S2 and S3 slope as seen in **Figure 2** reflected the free sliding situation where the safety coefficient was below 1 for 2 m blocks. For the stabilization case where the safety coefficient was over 1.35, the slip surface water parameters of the rock material on the slip surface were determined by slice weight analysis. Since the proposed analysis method was available for anchorage rock slopes, GEO 5 was directly used in this study. Because it was clear that completely degraded shale forming a weak slope over asphalt filled stabilized rock mass or completely free slip ground. GEO 5 method was preferred as the most suitable method for characterization of the free stability of the slope. RocScience software was also using finite element mesh programs and the parameters at this case of complete failure were determined. GEO5 analyzes were performed using the slice method. In this method, the safety coefficient is obtained by decreasing the shear strength parameters of the material forming the slope. GEO5 program calculated the 1.35 safety coefficient using shear force and resistive load. The reduced resistive load reduction method produced slide on slice weight principles. On the effective pore pressure, the rock failure by shear performs on below Eq (23).

$$\sigma'_1 = \sigma'_3 \cdot \frac{1 + \sin\varphi}{1 - \sin\varphi} + 2 \cdot c \cdot \frac{\cos\varphi}{1 - \sin\varphi} \tag{23}$$

As a result of analysis, shear resistive work performed in the field of geotechnical stabilization, the future should be considered a danger to very large fills and the filling cracked field should be determined according to the method of stabilization. Also within the project study area will be opened due to urban use preventive methods to investigate the instability in the region and it is important to develop a separate.

The stability was increased by compost rock cracks filled with asphalt/fly ash density reducing the water sorption content of uniaxial test blocks for 25 volume % rock cavity by 85% asphalt and 15% fly ash filling as shown in **Figure 15**.

The uniaxial compression strength of shale was increased to over 9 MPa by compost rock cracks filled with asphalt/fly ash density reducing the water sorption content of uniaxial test blocks for 25 volume % rock cavity by 85% asphalt and 15% fly ash filling as shown in **Figure 16**.

The uniaxial compression strength of shale was decreased to lower 8 MPa by compost rock cracks filled with increased fly ash content increasing the water sorption content. The uniaxial compression strength decreased to lower values in

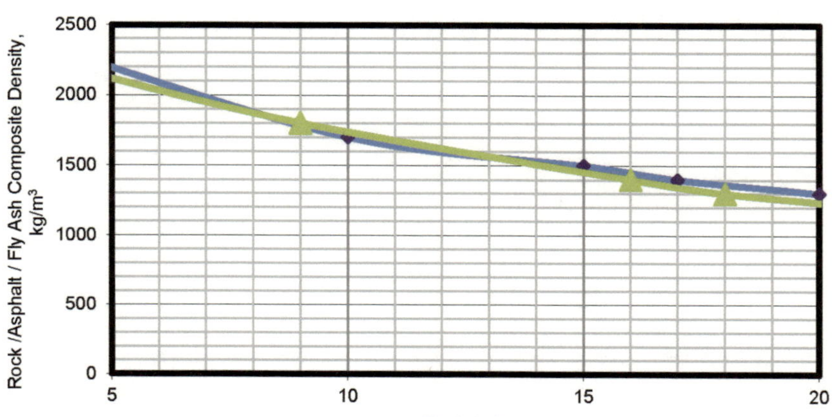

Figure 15.
The Bulk Density of the asphalt/fly ash filled Rock composite regarding Fly Ash Addition Vol%.

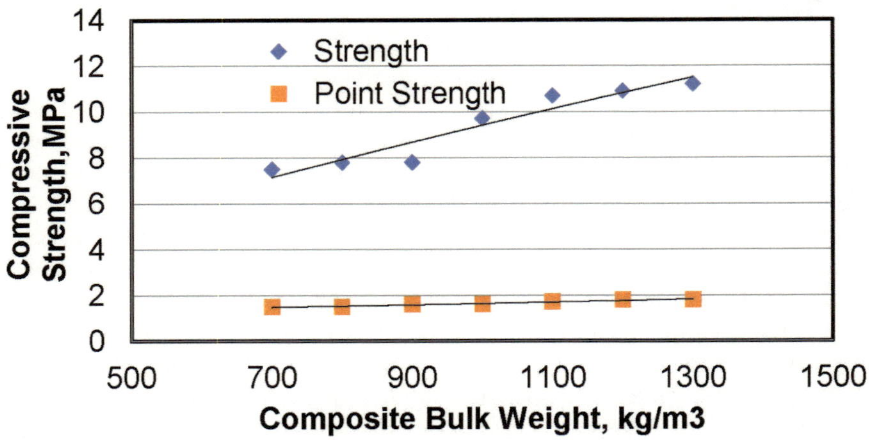

Figure 16.
The Compression Strength and I_s of the asphalt/fly ash filled Rock composite regarding Bulk Weight.

tested blocks for 25 volume % rock cavity by 80% asphalt and 20% fly ash filling as shown in **Figure 17**.

The water saturation of shale tested blocks was decreased over 70% by compost rock cracks filled with increased asphalt content over 60 vol % with increasing the cohesive filler content in tested shale blocks for 25 volume % rock cavity as shown in **Figure 17**.

In the scope of this study, both the numerical analysis and the proposed new asphalt flay ash fillings were evaluated for the reinforcement of free slides and stabilization of the migrating slope excavation, as well as the necessary weight slice analyses by GEO5 to ensure slope stability in the case sections of Avgamasya Pit Quarry No 2. The asphalt filling performance for free rock sliding was managed for slopes S1 S2 and other critical sub water perched face as seen in **Figure 3** caused water filled cracks and weak sub face soil texture.

5.1 Stabilization work and free sliding land and asphalt crack fill study

As a free slide soil -shale slip type was driven possible in rock falls of 2–5 m facing slopes, the main concepts of asphalt flay ash mixture filling was considered for stability and reinforcing weight slices over analyzed slip surface as explained above. Here, the rock cohesion was improved as an isolated block between water

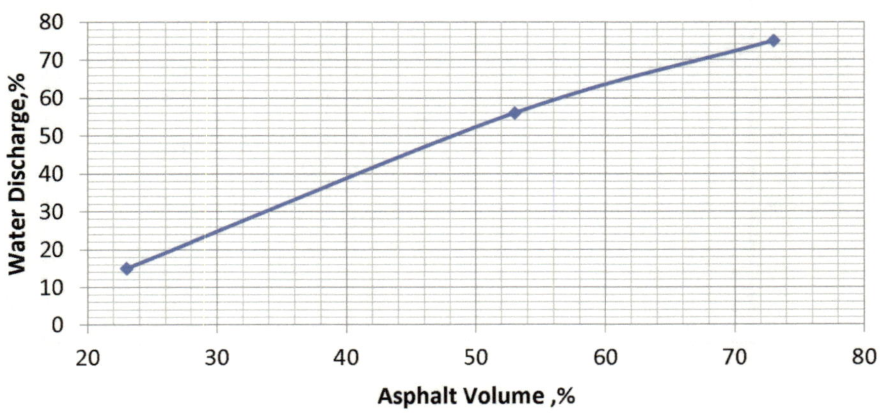

Figure 17.
The Water discharge of the asphalt/fly ash filled Rock composite regarding Asphalt Weight.

planes. The critical slopes were investigated that this block will not slip on planes with certain friction coefficients by improved cohesive matter of asphalt bound filling. The most important part of the problem was to determine the numerical values of crack filling performance by asphalt ash mixture that characterize the region in the stabilization analysis based on this theoretical idea. Numerical values of cohesion in stability problems determined by crack discontinuity. It can be summarized as the orientation of the surfaces, the average friction coefficients between these surfaces, the dimensions of the sliding wedge and the crack water pressures between the surfaces. Since these asphalt bound composite rocks were tested in various measuring techniques, the stability analysis of a rock slope should be done within the maximum and minimum cohesion values of these binder compost properties. The smallest of the safety numbers to be obtained should be used in the reference sizing, by the extensometer wire measurements as seen in **Figure 18**.

The rock fall, 3 m length crack elevation, asphalt compost filling made difference between the top and the heel point on 15 m. 30 m maximum height of the slope. The frequency distributions of the discontinuity orientations in the region are obtained in the form of a contour diagram as a result of placing a large number of discontinuity orientation measurements in a co-area network and a certain statistical evaluation. From this diagram, the maximum frequency orientations are called the dominant discontinuity orientations and they determine the planes to be used in stability analysis (Eq. (7)).

The mechanical properties of these discontinuity planes in terms of the stability of the rock slopes were free failures and friction coefficients. These properties can be determined as a result of shear tests of samples with discontinuity taken from the field. The Friction matter was critical in shale slopes due to low friction angle of below 22°. The force required to slide a block weighing W on a horizontal plane in a direction parallel to the plane must exceed the co-failure and friction force between the two surfaces. If the cohesion between two surfaces is assumed to be zero, the force required for shear should be W tgϕ,n weight slice chart. (**Figure 19**). The shear driving force reduced by strength planar levels auger bored and asphalt fill - wire net anchorage were applied as seen in **Figure 19**.

The free slip motion, shear forces in a slope was varied depending on the volume of the mass that is able to slide. Accordingly, the shape and therefore the weight of the mass that can slide should be found. In order to perform this procedure, it is necessary to know the properties of the mass that can slide and to be included in one

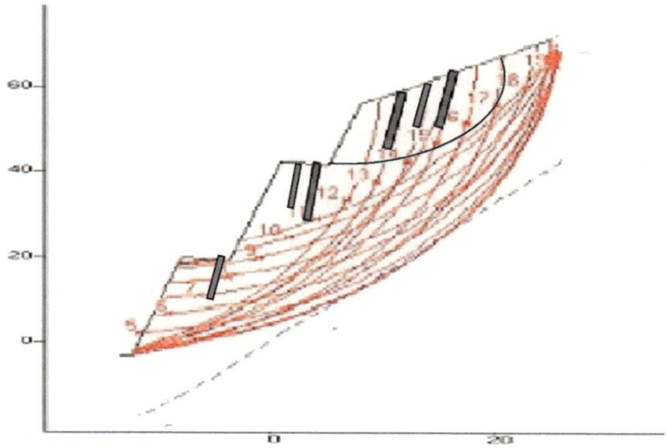

Figure 18.
Asphalt Crack Landfill Application cross-section on hazardous free slip area.

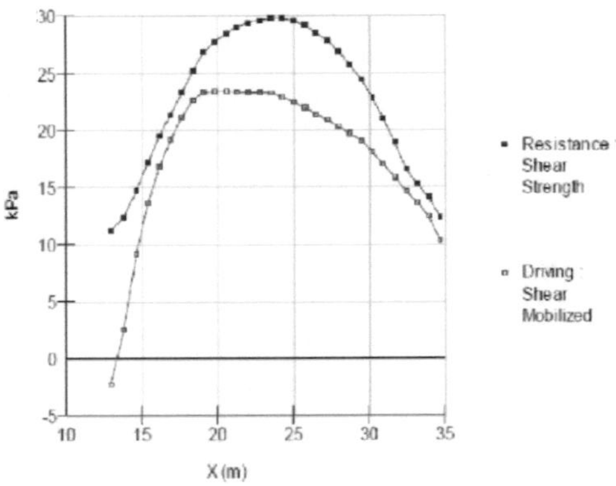

Figure 19.
Free slide shear load and driving forse over slice of slip surface on the base of water saturation of rock/ AsphaltComposite.

of the rock or ground class. Since the soil rock facing slip grounds was considered as heterogeneous, continuous and mostly anisotropic, and the mechanical properties of rocks, which had a discontinuous environment, and therefore the slice method was considered to be applied in even those free rock falls. It was impossible to apply the stability analysis model methods for rock falls applied to the open pit mines and road slopes of the heavily cracked rock slopes with the thought that the material itself was cut during the slope deterioration. Although The free soil slide was occurring with weight slice Bishop methods at 50–60 m length, the rock free slide was calculated for this shale formation so that a vertical limestone slope was fallen down at a height of 7–8 m That was the most common case to be broken as well in the past.

Although the idea of rock blocks sliding on a plane in rock slopes is an element, it is seen that rock wedges (weight blocks) bounded by two or more surfaces are formed as a result of the intersection of various discontinuities in nature, and their stability is more common in engineering works. In such cases, the rock wedge can slide, not on a plane, but on two planes, and its movement can be in the direction of the intersection of two planes. If it is, the event may occur.

The asphalt composed landfill of cable anchorage slopes developed cohesive resistivity and decreased perched water level on free sliding surfaces and the safety factor values reached over 1.5 and 1.8.

6. Conclusions

Asphaltite open pit mine and asphaltite excavation seam was hard to control against free sliding rock surrounding areas. There were 60 m even higher steeper slopes at angles over 65°. There were sliding large land marly shale masses or shattered shale rock formations. Underground water and harsh climatic conditions contain high risk hazard areas in operation site with higher risk factor of free slide. In order to eliminate rock falls and related events, significant precautions should be taken. The rock fall risk may ease to take precautions using asphalt composite filling. Even the application of elimination rock falls by wire net may reduce water content of soil. In this research, the pillar of 3 m and wire net were used with stabilization. The stability mixture of asphalt fill at certain thickness of 20 mm increased the safety. The low strength and porosity was critical for slide. The hazardous area could be safer by cohesive asphalt bound of rock cracks and long free rock slip surfaces out of water pressure.

Rock samples performed on the laboratory test results in the slope material not permeable that the cohesion value of 54–184.7 kPa, angle of internal friction of the 30.5–32.4° varied among marly shale/limestone with standard classifications. Stability analysis performed in the models srowed in **Figure 5** were unstable free slide rock interfaces. By use asphalt stabilization and pillar the hazardous free sliding slopes were converted to the stable condition. Shear force causing free slide was redueed by asphalt flyash mixture filling by cohesion of 400 kPa and 2 MPa shear strength with rock at 19 MPa uniaxial strength.

High strength landfill mass improved the slope stability. However, low strength foam concrete landfill might result in higher water discharge and drier soil condition. The pillar bottom layers avoided water saturation of cracks and slip surface bottom area even sequential top surfaces of slopes.

Dissociation detailed inclinometer observations provided in rocks on free slide area also offered a positive contribution to stability problems.

For those reasons, asphalt composite crack fill can use virtual any classical slope stability programs, rock slope stability calculations in order to do construction safety and dry soil matters as given factor over 1.35.

Symbols

c 'kg/cm^2	Effective Cohesion
c kg/cm^2	Cohesion
Φ'o	Effective internal friction angle
Φo	Internal friction angle
τ kg/cm^2	Shear stress
σ kg/cm^2	Normal stress
Ip	Plasticity index
Ll	Liquid limit
Pl	Plastic limit
Wopt	Optimum water content
γNatural g/cm^3	Natural unit volume weight
Saturated g/cm^3	Saturated unit volume weight
γDry g/cm^3	Dry unit volume weight
γkmax g/cm^3	Maximum dry unit volume weight
γs g/cm^3	Grain unit volume weight
k	Permeability coefficient
S1, S2, S3, S4	Şırnak free slide slopes no. 1, 2, 3, 4
S11	Sample taken from Şırnak Asphaltite Open Pit Mine free slide slope no.
SP	In the combined ground classification; poorly graded sand, gravel sand, fine material no
SW	In combined ground classification; well-graded sand, gravelly sand, little or no fine material
SC	In the combined ground classification; claystone sand, sand-clay mixture
GW	In combined ground classification; well graded gravel, gravel-sand mixture, fine material little or no
J_i	Joint density sequence at slice i
N	Serial distribution of Discontinuity
R_c	Shear risk factor
F	Safety factor
σ	Compression Stress

$\tau_{\theta i}$	Shear Stress at slip surface at slice i
u	pore presssure

Author details

Yildırım İsmail Tosun
Mining Engineering Department, Engineering Faculty, Şirnak University, Şırnak, Turkey

*Address all correspondence to: yildirimismailtosun@gmail.com

References

[1] Airey, G.D. and Collop, A.C., 2014, Mechanical and structural assessment of laboratory- and field-compacted asphalt mixtures International Journal of Pavement Engineering *Volume 17, 2016 - Issue 1: Asphalt compaction*

[2] Tashman, L., Wang, L. B. and Thyagarajan, S. 2007. Microstructure characterization for modeling HMA behavior using imaging technology. *Journal of Road Materials and Pavement Design*, 8(2): 207–238.

[3] Bieniawski, Z.T. 1967. Mechanism of brittle failure of rock Part I - Theory of fracture process. *I. J. of Rock Mech. and Min. Sc. and Geomech. Abstr.* 4: 4, s. 395-406

[4] Hoek, E. and Bray, JW, 1977, Rock slope engineering, Stephen Austin and Sons, Ltd., Hertford, 402 p.

[5] Lamp, W. T. and Whitman, RV, 1969, Soil Mechanics, John Wiley and Sons, New York

[6] Bishop, AW, 1955, The use of the slip circle in the stability analysis of earth slopes, Geotechnique, Vol. 5, 7-17.

[7] Hoek, E., 1970, Estimating the stability of excavated slopes in Opencast mines, Institution of Mining and Metallurgy, A105, A132

[8] Paşamehmetoğlu, LV, Özgenoğlu, A., Watermelon, C, 1991, Rock Slope Stability, 2 Print, T.M.M.O.B Mining Eng. Room Publications, Ankara, May.

[9] Anbalagan, R., 1992, Landslide hazard evaluation and zonation mapping in mountainous terrain, "Engineering Geology, 32:269-277,

[10] Fall, H, 1987, "Geotechnical Developments", DSI slope of the Stability and Retaining Structures Seminar, Samsun

[11] Anonymous, 2011, Şırnak Provincial Administration Reports.

[12] Anonymous, c 2011, "Turkey Earthquake Zone Map", Disaster and Emergency Management Bureau, Earthquake Department Ankara

[13] ASTM D616-07, 2007.Standard test method for Fly ash materials and compositions, slag materials, ASTM, Pennsylvania

[14] ASTM C 330, 2013.Standart Specifications For Lightweight Aggregates for Structural Concrete, ASTM, Philadelphia.

[15] ASTM D6024-07, 2007.Standard test method for Ball Drop on Controlled Low Strength Materials, ASTM, Pennsylvania.

[16] ASTM C 136, 2013.Standard test method for Sieve Analysis of Fine and Coarse Aggregates, Pennsylvania.

[17] Anonymous, 2013, GEO5 programs - Engineering Manuals - Part 1 - Part 2 http://www.finesoftware.eu/geotechnical-software/

[18] Anonymous, 2009, GEO5 programs - FEM - Theoretical Guide http://www.finesoftware.eu/geotechnical-software/

[19] Görög P & Török, A, 2007 Slope stability assessment of weathered clay by using field data and computer modeling: a case study from Budapest, Natural Hazards and Earth System Sciences, 7, 417-422, www.natu-hazards-earth-syst- sci.net

[20] Görög P, 2006, Stability problems of abandoned clay pits in Budapest, IAEG 2006, Paper number 295, The Geological Society of London

[21] Dramis, F., Sorriso-Valvo, M., 1994 "Deep-Seated gravitational slope

Deformations, related landslides and tectonics", Engineering Geology 38, 231-243,.

[22] Hoek, E., 2013. Practical Rock Engineering, notes by Evert Hoek Hoek. My http://www.rocscience.co

[23] Hutchinson, JN., 1995 "Landslide hazard assessment. Keynote paper. In: Bell DH (ed) landslides, Proceedings of 6th International Symposium on landslides ", Christchurch, New Zealand, vol 1 Balkema, Rotterdam, pp. 1805-1841,

[24] Kilic, R., Ulamış, K.., "Gölbasi (Ankara) Investigation of Mass Movements in the South," Bulletin of Engineering Geology, Engineering Geology Of Bulletin, Issue: 20 (75-86).

[25] Prusa, J., 2009, Comparison of geotechnic softwares - Geo FEM, Plaxis, Z-Soil, XIII to ECSMG, Vanicek et al. (Eds). cgts, Prague, ISBN 80-86769-01-1, (Vol. 2)

[26] Vaneckov, V, Laura J, J Prus, 2011, Sheeting Wall Analysis by the Method of Dependent Pressures, Geotec Hanoi - ISBN 978 - 604-82 - 000 - 8 ID No. / Pp. 7

[27] Wiley, L., 1987 "Stability Slope and Geomorphology Geotechnical Engineering", England,

[28] Tosun, Y. İ., Cevizci, H., Ceylan, H., 2014, Landfill Design for Reclamation of Şırnak Coal Mine Dumps - Shalefill Stability and Risk Assessment, ICMEMT 2014, 11-12 July 2014, Prag, Chekoslovakia.

[29] Tosun, Y. İ., 2014, A case study on use of foam concrete landfill on landslide hazardous area in Şırnak City Province, XX Congress of the Carpathian Balkan Geological Association, Tirana, Albania, 24-26 September 2014.

[30] Tosun, Y. İ., 2014, *Shale stone and Fly ash Landfill Use in Land-slide Hazardous Area in Sirnak City with Foam Concrete*, GM Geomaterials Journal, Vol 4, issue 4, pp 141-150 DOI: 10.4236/gm.2014.44014

[31] Yıldırım İ.Tosun, 2016, Kalker, Marn ve Şeylin Sünme Karakterizasyonu - Bitümlü Gözenekli Agrega için Don - Mikrodalga Kurutma-Bilya Darbe Dayanım Testi ile Sünme Etüdü, AGGRE 2016, 8th Internatıonal Aggregates **Sy**mposıum, October 5-7, Istanbul, Turkey

[32] Tosun, Y.İ., 2016, Use of Modified Freeze-Drop Ball Test for Investigation the Crack Propagation Rate in Coal Mining- Case Study for the Şırnak Asphaltite Shale, Marly Shale and Marl in Şırnak Coal Site, *IBSMTS 2016, 1th Internatıonal Black Sea Symposium on Mininig and Tunnelling*, November 02-04, Trabzon, Turkey.

[33] Moulia, M., Khelafib H., 2008. Performance Characteristics Of Lightweight Aggregate Concrete Containing Natural Pozzolan, *Building and Environment* , 43, s. 31–36

[34] Park, C. K. Noh, M. H. Park, T. H. , 2005. Rheological Properties Of Cementitious Materials Containing Mineral Admixtures, *Cement And Concrete Researchs* 842-849

[35] Olard, F. and Perraton, D. 2010. On the optimization of the aggregate packing characteristics for the design of high-performance asphalt concretes. *Road Materials and Pavements Design*, 11 (Suppl. 1): 145–169. doi:10.3166/rmpd.11hs.145-169

[36] Tashman, L., Masad, E., Peterson, B. and Saleh, H. 2001. Internal structure analysis of asphalt mixes to improve the simulation of superpave gyratory compaction to field conditions. *Journal of the Association of Asphalt Paving Technologists*, 70: 605–645.

[37] Török, Á. Bögöly, G., Czinder, B.. Görög, P, Kleb, B.& Vásárhelyi, B., Lovas, T. Barsi, Á., Molnár, B., Koppányi Z. & Somogyi, J. Á., 2016, Terrestrial laser scanner aided survey and stability analyses of rhyolite tuff cliff faces with potential rock-fall hazards, an example from Hungary, Eurock 2016, Cappadicia, Volume: 877-881

[38] Á Török, Á Barsi, G Bögöly, T Lovas, Á Somogyi, and P Görög, 2017, Slope stability and rock fall hazard assessment of volcanic tuffs using RPAS and TLS with 2D FEM slope modelling, Nat. Hazards Earth Syst. Sci. Discuss., doi:10.5194/nhess-2017-56, 2017

Chapter 6

The Potential of Remote Sensing to Assess Conditioning Factors for Landslide Detection at a Regional Scale: The Case in Southeastern Colombia

Nixon Alexander Correa-Muñoz and Carol Andrea Murillo-Feo

Abstract

This landslide detection research applied remote sensing techniques. Morphometry to derive both DEM terrain parameters and land use variables. SAR interferometry (InSAR) for showing that InSAR coherence and InSAR displacement obtained with SRTM DEM 30 m resolution were strongly related to landslides. InSAR coherence values from 0.43 to 0.66 had a high association with landslides. PS-InSAR allowed to estimate terrain velocities in the satellite line-of-sight (LOS) in the range − 10 to 10 mm/year concerning extremely slow landslide displacement rates. SAR polarimetry (PolSAR) was used over L-band UAVSAR quad-pol data, obtaining the scattering mechanism of volume and surface retrodispersion more associated with landslides. The optical remote sensing with a multitemporal approach for change detection by multi-year Landsat (5, 7 and 8)-NDVI, showed that NDVI related to landslides had values between 0.42 and 0.72. All the information was combined into a multidimensional grid product and crossed with training data containing a Colombian Geologic Service (CGS) landslide inventory. A detection model was implemented using the Random Forest supervised method relating the training sample of landslides with multidimensional explanatory variables. A test sample with a proportion of 70:30 allowed to find the accuracy of detection of about 70.8% for slides type.

Keywords: landslide detection, morphometry, multi-InSAR, pol-SAR, NDVI, Random Forest

1. Introduction

This research aimed to design a model for the detection of landslides at the regional scale using Earth Observation data and remote sensing techniques. The following techniques were used to achieve this goal.

Landslide inventory (LI), for example the CGS-SIMMA[1] as the base for the evaluation of the multivariate data generated by new technologies. It provides information

[1] Sistema de Información de Movimientos en Masa del Servicio Geológico de Colombia; in English: Mass Movement Information System from the Colombian Geologic Service.

about the vector representation such as point objects, date, municipality, mass distribution, type, sub-type, material and origin contributing to landslide detection [1].

At regional scale, landslide detection and landslide distribution analysis allows to analyse the distribution and classification of landslides [2]. These analysis use univariate and multivariate statistical methods, obtaining weights by the correlation between landslides occurrences and conditioning factors. Weight of Evidence (WofE) method is a bivariate approach where the landslides inventory is used to calculate weights of conditioning factors in order to delineate potential areas of landslides. Logistic regression (LR) is one of the statistical methods more used to evaluate the relationship between landslides and related factors [3].

Remote sensing techniques have been used for landslides detection [4]. One of them is morphometry based on Digital Elevation Models (DEMs), for a quantitative analysis of the land surface topography. In this way, the geomorphometry analysis derives land surface parameters such as slope, aspect, curvature or basic local descriptors, regional parameters as catchment area, parameters connected with hydrology like topographic wetness index and so on [5]. Terrain parameters can be related to landslides to build detection models [6]. Change detection technique [7] based on the sudden disappearance of vegetation by NDVI difference computation can serve to landslide detection. Also, SAR-based techniques in rural areas can help to landslide detection [8].

Interferometric spaceborne radar to measure linear deformation rates has been implemented on several studies to landslides detection [9]. PS-InSAR (persistent scatterers) allows measurements with millimetre accuracy of individual features. PS-InSAR is differential measurements with respect to a reference point that is assumed to be stable [10]. PolSAR imagery allows to characterise objects on the ground based on that different structures and geometries show different backscatter values at different SAR polarisations. Quad polarimetric SAR data content information to landslide detection in forested areas under the assumption that the dominant mechanism is surface scattering with high homogeneity [11].

This way Earth Observation (EO) data encompasses sensors like SAR, optical images and GPS on board of platforms either satellite-based, aircraft-based or ground-based provide high spatial, temporal, and spectral resolution to geohazard studies [12]. The above combined with machine learning (ML) techniques like Random Forest, allow the mapping, monitoring and modelling of landslides occurrences.

2. Landslide inventory and earth observation data

In this section we describe the study area, the landslide inventory, taken the from the Colombian Geologic Service (CGS), and the Earth Observation data.

2.1 Study area

The study area is located at the southwest of Colombia and covers the inter and central Andes Mountains in a rectangle within the following WGS84 system coordinates: 02°06′53.50″ North, 76°48′49.71″ West and 02°34′27.81″ North, 76°24′32.25″ West. In here, we found elevations between 848 and 4932 m.a.s.l. The area covers the inter-Andean valleys of Cauca river and the central mountain range of the Andes in the southwestern of Colombia. Former is to comprise Tertiary and Quaternary formations and above there are volcanic ash depositions. The Central Mountain Range of the Andes in Cauca state, have quaternary deposits at the summit and its western slope it is found diabase rock. **Figure 1** shows the study area on Alos Pasar elevation data as background and the sheets of the topo-map provided by the National Geographical Institute 'Agustin Codazzi' (IGAC).

2.2 Landslide inventory

The CGS-SIMMA geo-service allowed to build the landslide inventory database for training the detection model. Landslide database contains an inventory of

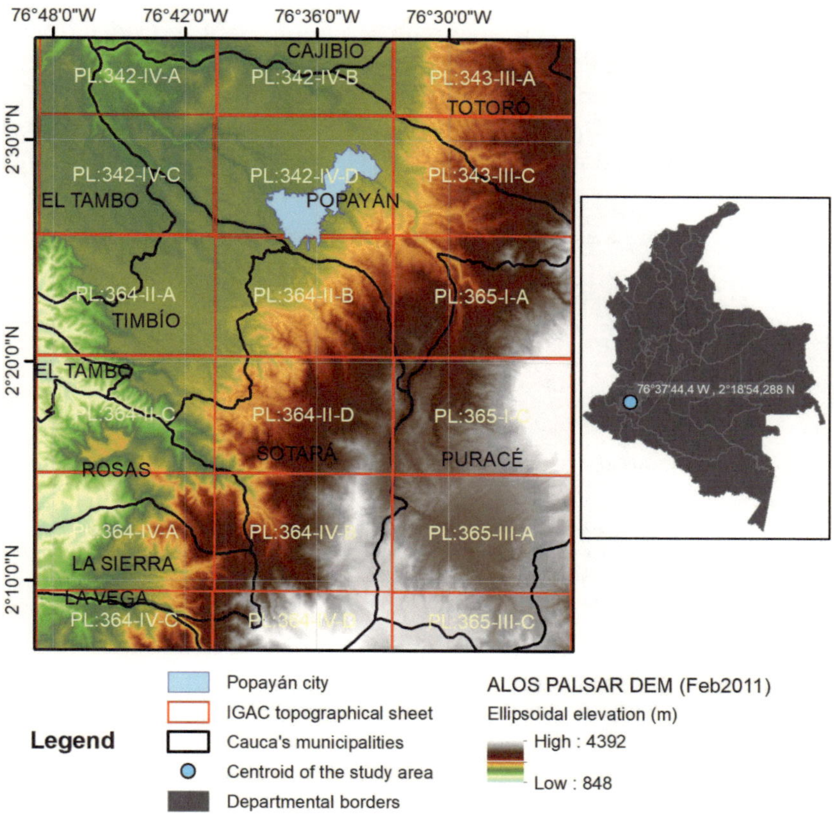

Figure 1.
National and regional location of the study area.

LS Type	Detritus	Rock	Earth	Other	Total	%
Slides	77	14	78	9	178	77.4
Falls	15	8	14	1	38	16.5
Creep	0	2	4	1	7	3.0
Flows	0	0	3	3	6	2.6
Lateral Spreads	0	0	1	0	1	0.4
Total	92	24	100	14	230	100
%	40.0	10.4	43.4	6.1	100	

Table 1.
CGS-SIMMA landslide frequency distribution.

geomorphologic and catalogue type. The former contained more precise information in its thematic attributes. The overall distribution of SIMMA landslide inventory was 77.4% of slide type, 16.5% of fall type, and 6.1% of flow and creep type for a total of 230 registered events (**Table 1**). The spatial information contained only allowed mapping in a shapefile of point type.

2.3 Earth observation data

Table 2 summarise the EO data used in this research. EO data corresponds to DEMs, optical remote sensing and radar remote sensing. DEMs provided from SRTM DEM 30 m resolution and Palsar RTC elevation data at 12.5 m. Radar data had as source the spaceborne-based ESA-Copernicus Sentinel-1 satellites and the aerial-based UAVSAR platform. Optical data was provided by time series analysis of the multi-year Landsat 5, Landsat 7, and Landsat 8 NDVI surface reflectance images.

Data	Platform or mission	Band	Resolution (m)	Cycle (days)	Wide swath (km)
DEM	SRTM/NASA[1]	C/X	30	—	225
DEM	AP-RTC/JAXA[2]	L	12.5	46	70
Dual-pol (VV/VH)	Sentinel-1/ESA[3]	C	15	24	250
Quad-pol	UAVSAR/NASA-JPL[4]	L	7	365 days	16
Optical	Landsat 8 SR/NASA[5]		30	16 days	185

[1]*Retrieved from http://srtm.csi.cgiar.org/ and https://earthexplorer.usgs.gov.*
[2]*Retrieved from https://www.asf.alaska.edu/doi/105067/z97hfcnkr6va/.*
[3]*Copernicus Sentinel data 2014. Retrieved from ASF DAAC 29 April 2017, processed by ESA.*
[4]*UAVSARdatacourtesyNASA/JPL-Caltech, retrieved from https://uavsar.jpl.nasa.gov/cgi-bin/data.pl.*
[5]*Retrieved from https://search.earthdata.nasa.gov.*

Table 2.
Earth observation data used in this research.

3. Methodology

The main remote sensing techniques over Earth Observation data to extract features and conditioning factors related to landslides are listed in **Table 3**.

Figure 2 shows the functional representation of the EO data indicating the entire flow processing needed to develop the detection model from multi-dimensional

Approach	Variable or method	Software	Software type
Morphometry	Land surface parameters	Demanal/SAGA/R/ArcSDM[1]	Open-source
Spaceborne-based InSAR	InSAR measurements (coherence and displacement)	SNAP toolbox/ SARProZ[2]	Open-source and Commercial
Multi-InSAR	Deformation velocities of persistent scatterers	SARProZ[3]	Commercial
Aerial-based PolSAR	Surface, volume, and double-bounce scattering mechanism	PolSARpro_v5.0[4]	Open-source
Optical remote sensing	NDVI vegetation indices	Google Earth Engine[5]	Open-source
Binary model	WofE analysis and Logistic regression	ArcSDM[6]	Open-source

Approach	Variable or method	Software	Software type
Multidimensional data fusion	Random Forest supervised method	QGIS/ArcCatalog/R software[7]	Open-source and commercial

[1]DEMANAL package of BLUH software [13].
[2]Retrieved from https://step.esa.int/main/download/snap-download/.
[3]SarProZ software retrieved from www.sarproz.com.
[4]ESA's PolSARpro-v5.0 software.
[5]Retrieved from https://earthexplorer.google.org.
[6]Arc-SDM tool [14].
[7]R software [15].

Table 3.
Methods and approaches of remote sensing techniques used in this research.

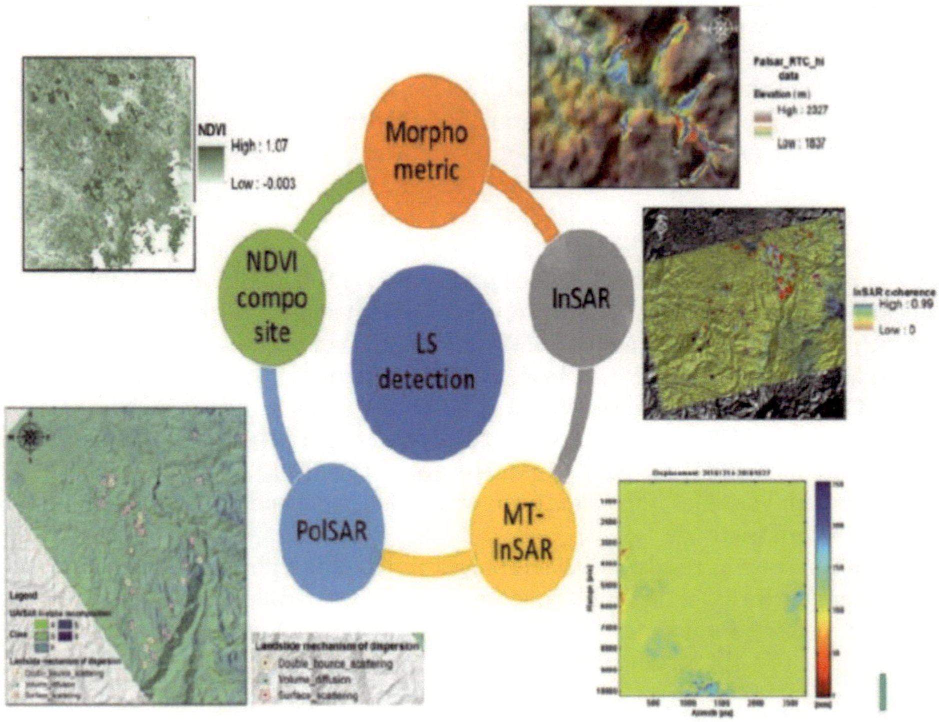

Figure 2.
Functional data for model building.

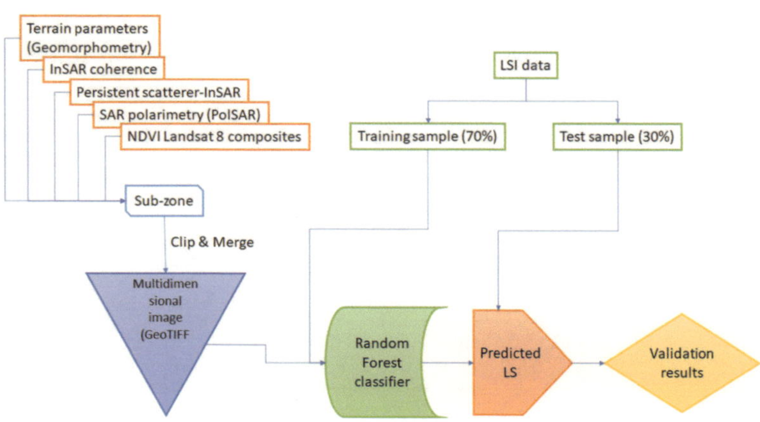

Figure 3.
Chain of processing to landslide detection model.

data, data training and test of landslides. Also, the scheme shown in the **Figure 3** indicates the fusion of all geospatial information by the Random Forest method.

The CGS-SIMMA landslide inventory was split into training and test subset in a proportion of 70:30 in concordance with the study made by Huang and Zhao [16] in order to determine the accuracy of each remote sensing method applied and the detection model generated in this research.

4. Results

The results of remote sensing techniques implemented in the study area are presented in this section.

4.1 Land surface parameters (LSP) related to landslides

The DEMs Palsar RTC elevation data, SRTM DEM 30 m and 90 resolution, and ASTER-GDEM were evaluated in relation to GPS control points and a reference Topo DEM obtained by interpolation of contours at a scale of 1:25 K. The best in terms of vertical root-mean-square-error (RMSE) was SRTM 30 m resolution, and its accuracy at a 95% confidence level (7.8 m) corresponded to a better scale than 1:25 K. ALOSP RTC elevation data was the second best DEM. For this reason we derived from the latter the land surface parameters (LSP) at 12.5 m resolution using algorithms implemented on SAGA software. **Table 4** shows the results of a vertical accuracy assessment of the global DEMs used in this research.

The land surface parameters: slope, aspect, curvature [17], topographic wetness index (TWI) [18], valley depth (Vdepth) [18], convergence index (CONVI) [19], flow path length (FPL), and insolation [20], were converted into independent components by using Principal Component Analysis (PCA) [21]. These were used as independent variables into a landform detection model and a landslide regression model by WofE methods.

Table 5 shows the results of WofE analysis to relate morphometric and land use conditioning factors with landslide inventory. Only the variable with its class with the most studentised contrast (bigger than 2) C/s(C) is shown. **Figure 4** shows the unit soils at a scale of 1:100 k, which covers plain, undulated and mountainous terrains.

4.2 InSAR measurements

InSAR measures are phase, coherence, and displacement and they are obtained by the cross-correlation between two or more SAR images to process the line-of-sight displacements. This research used C-band data provided by Sentinel/1 ESA's Copernicus programme. In this investigation, the effect of a DEM on the InSAR

Metric	ALOSP	ASTER	SRTM1 (30 m)	SRTM3 (90 m)
RMSE (m)	39.68	43.52	19.95	22.77
SZ (without bias) (m)	4.427	5.331	3.937	4.595
NMAD (m)	3.806	4.663	3.093	3.883
LE95 (without bias) (m)	8.853	10.236	7.795	9.144
SZ=	4.62 + 80.5*tan (slp)	5.60 + 89.8*tan (slp)	3.54 + 109.9*tan (slp)	4.689 + 100*tan (slp)

Table 4.
Vertical accuracy of the global DEMs compared with Topo-map DEM reference.

LSP	The range of study area	Class	C/s(C)
Slope	0° - 79.4°	6.6° to 19.9°	2.1
		13.2° to 26.5°	2.9
FPL	0–2598 m	0–371 m	2
Soil unit	9 classes	Humid hill lands (LQ) and very wet cold mountain (MK)	3.5
Landform	12 classes	Backslopes	3.6

Table 5.
WofE studentised contrast > 1.5 for LSP and land use conditioning factors.

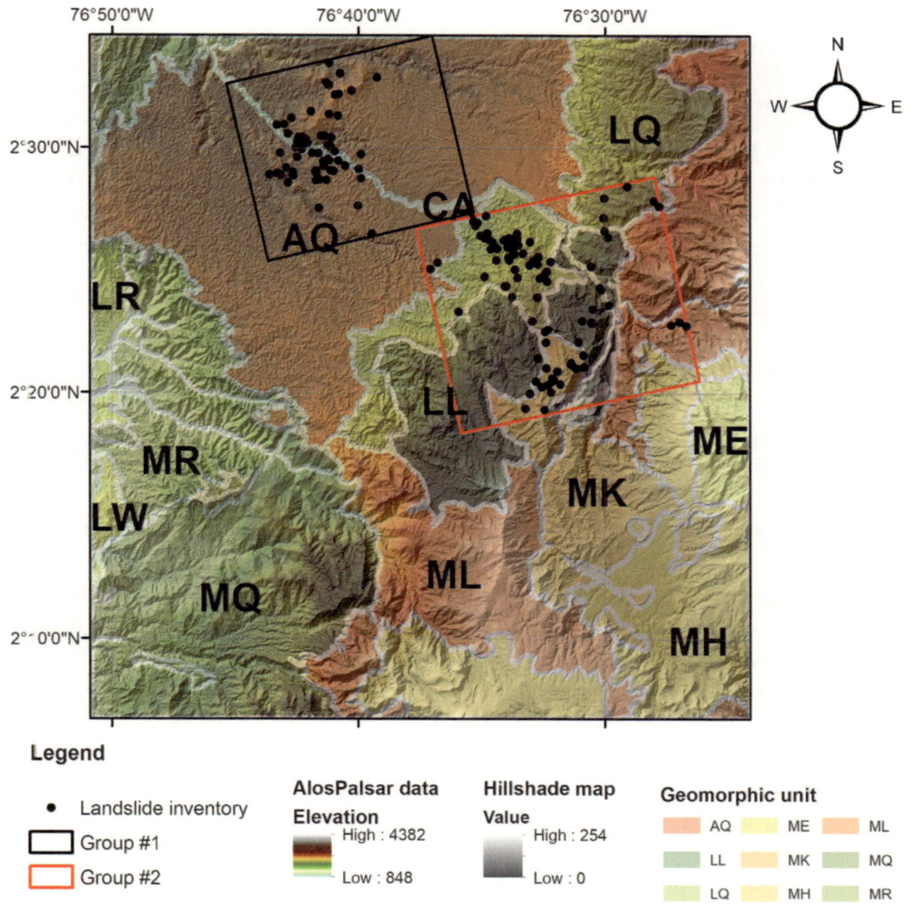

Figure 4.
Soil units of the study area.

processing was determined [22], concluding that DEM did not have an effect on InSAR coherence but if the InSAR phase is unwrapped with the DEM variable there are significance differences. The above is due to the inaccuracies of external DEM.

Two landslide regression models obtained either with InSAR coherence or InSAR displacement from a DEM variable showed that SRTM DEM 30 m resolution had the highest association with landslides inventory (**Table 6**). These models had an accuracy of 62% and 68% respectively.

From the point of view of the InSAR coherence measurement, all DEMs are not statistically significant. However, Topo-map showed a weak relation to landslides.

DEM	InSAR coherence			InSAR displacement		
	Coeff.	Pr(> z)	Pr(> Chi)	Coeff	Pr(> z)	Pr(> Chi)
Intercept	−9.345	< 2e-16***		−13.15	< 2e-16***	
SRTM DEM 30 m	2.627	0.853	0.0014**	0.870	0.0003***	1.56e-9 ***
SRTM DEM 90 m	−2.674	0.850	0.816	−0.467	0.04*	0.22
Palsar RTC	−0.085	0.522	0.069	0.323	0.084.	0.057.
Topo-map	0.474	0.004**	0.0031**	−0.017	0.924	0.924
AUC	0.62			0.68		

*Significance of codes: 0 = ***, 0.001 = **, 0.01 = *, 0.05 =., 0.1 = ' '.*

Table 6.
Results of InSAR regression analysis by LR method.

Topo-map and SRTM 30 m contributed to a better explanation of the linear regression model. InSAR displacement with a DEM variable indicated that SRTM 30 m had the lowest p-value suggesting a strong association of the elevation with the probability of having a landslide with a positive coefficient. ANOVA verificated that by adding the SRTM 30 m to the regression model significantly reduces the residual deviance.

WofE analysis showed that the maximum studentised contrast for InSAR coherence was in the range of 0.43 to 0.66.

Multi-InSAR processing by PS-InSAR method allowed to estimate the deformation rate in relation to a reference point target which is assumed to be stable. This approach overloads the substantial limitation of InSAR measurements: the spatial and temporal decorrelation and atmospheric distortions due to ionospheric electron density and tropospheric water vapour.

The C-band sensor on-board the Sentinel-1 satellite served as input data to implement PSInSAR processing to estimate the annual linear velocities and the time series of deformations. **Table 7** indicates the geometrical characteristics of S1_A of the ESA' Copernicus used for PS-InSAR in the study area. The perpendicular baseline in all cases was lower than 150 m, which is an adequate value for studies of terrain deformation. The results were a deformation map which consisted in a set of selected points (12 pts./km^2) with both the information of the estimated LOS velocity (in the range − 10 mm/year to 10 mm/year) and the accumulated displacement. Ordinary Kriging (OK) method allowed to predict the LOS velocity on landslide inventory (**Figure 5**). A terrain displacement between −4.5 mm/year and 4.8 mm/year in ascending mode, and between −2.7 mm/year and 7.7 mm/year in descending mode was predicted with the satellite LOS velocity, indicating the movement towards (positive values) and away (negative values) the sensor respect to the master radar scene. The prediction variance found was lower than 1.6 (mm/year)^2 in ascending orbit, and lower than 5.63 (mm/year)^2 in descending pass.

Zone	Dates	Stack	Pass/Pol	Bn(m)	Bt(days)	IncAnc (°)
SE	10–2014/09–2015 and 01–2016/09–2016	24	Asc/VV	7 to 144	24–384	34.2
NW	10–2014/09–2015 and 01–2016/09–2016	21	Asc/VV	7 to 144	24–384	34.2
NE	10–2014/05–2016	21	Des/VV	2 to 117	24–312	34.1

Table 7.
Sentinel-1-IW-SLC dataset.

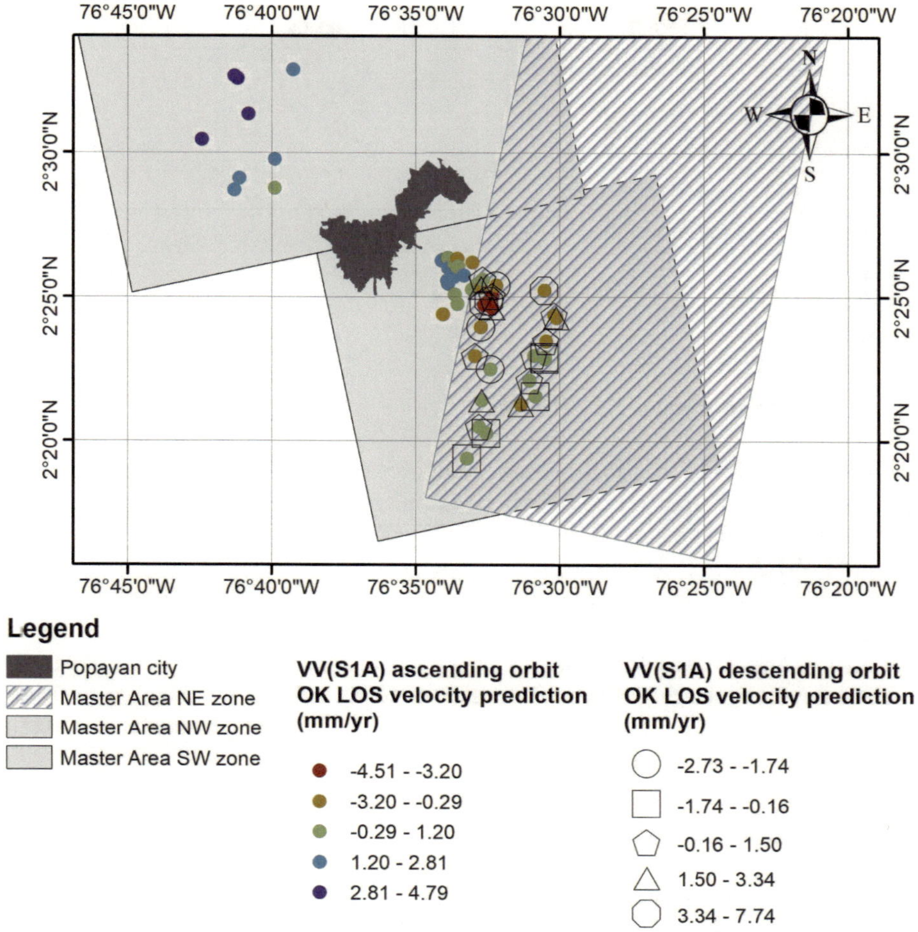

Figure 5.
OK prediction of the displacement velocity in mm/year.

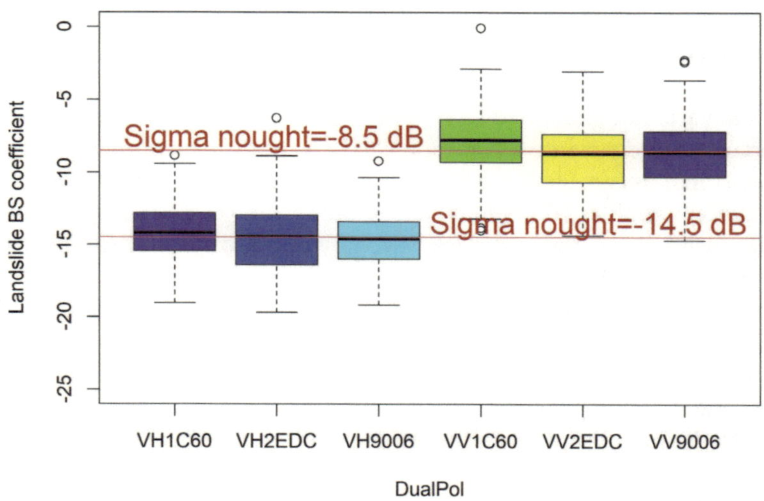

Figure 6.
BS-coefficient (σ°) in landslide inventory.

H-α category	Landslide frequency	Relative frequency
4-Medium entropy and multiple scattering	12	9.5
5-Medium entropy and volume scattering	63	50
6-Medium entropy and surface scattering	32	25.4
7-Low entropy and multiple scattering	15	11.9
8-Low entropy and volume scattering	4	3.2
Total:	126	100

Table 8.
H-α decomposition of quad-pol data.

4.3 PolSAR unsupervised classification

Dual Pol-Sentinel-1 analysis allowed to analyse the sigma nought scattering coefficient with VV and VH polarisation. Copolarization backscattering (−8.5 dB) was higher than cross-polarisation (−14.5 dB) (**Figure 6**).

Quad Pol-UAVSAR decomposition allowed to define the mechanism of scattering on landslides inventory using entropy/alpha within the Cloude Pottier method. The scattering mechanism dominant in the study area were volume scattering (vegetation) and surface scattering (**Table 8**). The results indicate that 50% of the landslide have a scattering mechanism of volume and 25.4% of surface type. The WofE method validated that the H-α classification of volume and surface scattering were highly related to landslides.

4.4 NDVI time series analysis

Time series analysis of the multi-year Landsat NDVI was used as input data for the change detection analysis. In the period 2012 to 2017, the multi-year Landsat NDVI cloud-free yearly composites through Google Earth Engine did not show the statistically significant trends in vegetation. But WofE analysis found that the NDVI range with the highest association to the landslide inventory was between 0.40 and 0.72. **Figure 7** shows that only the years 2012, 2014 and 2016 covers, without clouds, the landslide inventory distribution with median values of NDVI.

4.5 Model to the detection of landslides

All of the variables generated (25) in this research by remote sensing techniques were overlapped and cut into a common sub-zone and then combined into a multidimensional image. Here are found the classification variables. Then the effect of classification variables (derived from remote sensing techniques) over a target variable (landslide inventory) was measured by the algorithm of supervised pixel-based classification called Random Forest. Test data in a proportion of 30% of the entire data set allowed to obtain an independent validation. **Table 9** show the Random Forest classification with an overall accuracy of 70.8%. The user's accuracy refers to the correct classification of the type of movement in relation to the referenced one, and the producer's accuracy refers to the commission or inclusion error. Due to the high frequency of rotational and translational slide, the method was successful, which did not happen with other less frequent types.

Figure 8 shows the results of the Random Forest classification for the landslides types: debris fall, flow, planar translational, rotational and translational. Rotational and translational slides had a producer's accuracy of 80% and 91% respectively

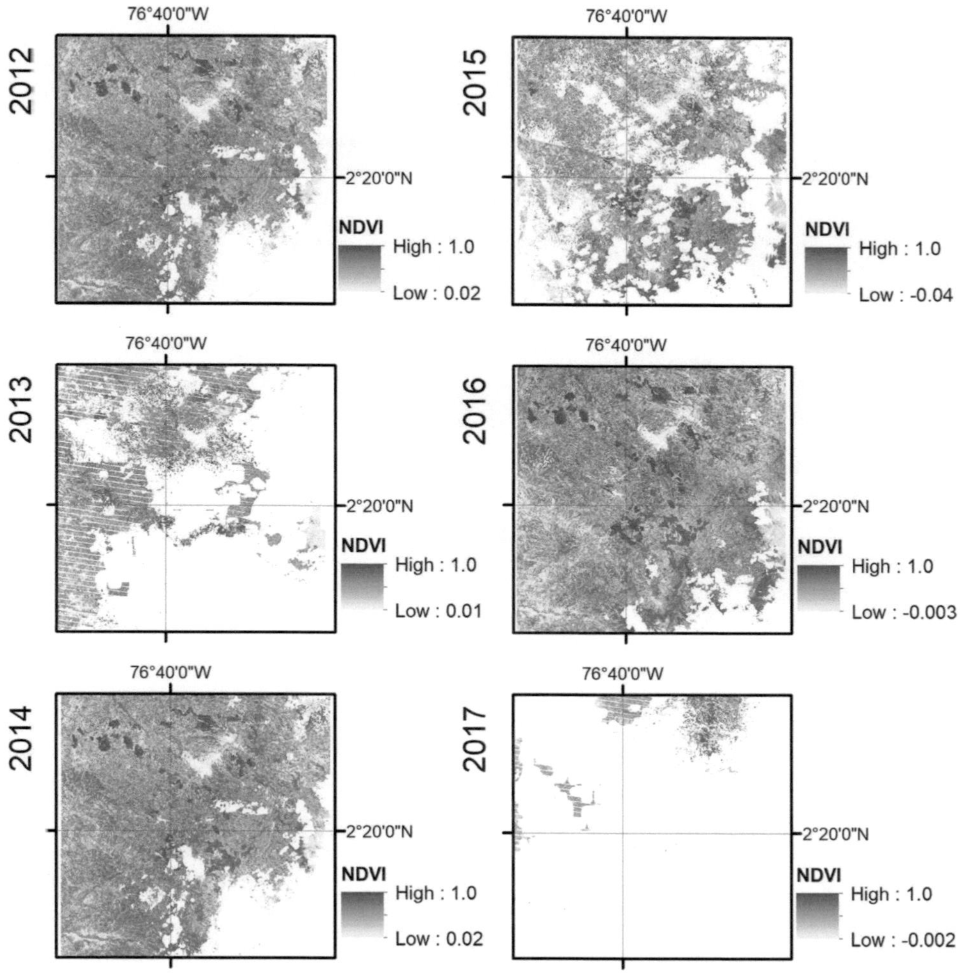

Figure 7.
7-year LANDSAT NDVI composites (2010 to 2017).

Random Forest classification	Topic	Sw
	Overall accuracy	70.8%
Training model	Rotational slide	66.7%
	Translational slide	96.2%
User's accuracy	Rotational slide	66.7%
	Translational slide	71.4 $
Producer's accuracy	Rotational slide	80%
	Translational slide	90.9%

Table 9.
Random Forest classification.

resulting in omissions errors of 20% and 9% for each one. User's accuracy for the same type of landslides was of 67% and 71% indicating commission errors of 33% and 29%. The overall accuracy was 70.8%.

Figure 9 shows the Mean Decrease Accuracy implemented in Random Forest. The variables which contributed more to the study were PolSAR, the displacement InSAR, the NDVI and the morphometric variable slope.

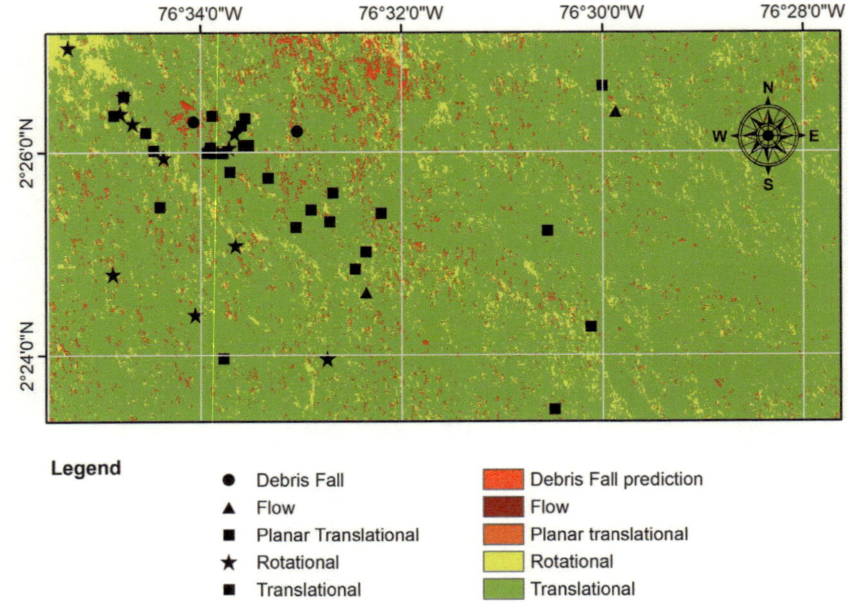

Figure 8.
Detection model of landslides by the Random Forest method.

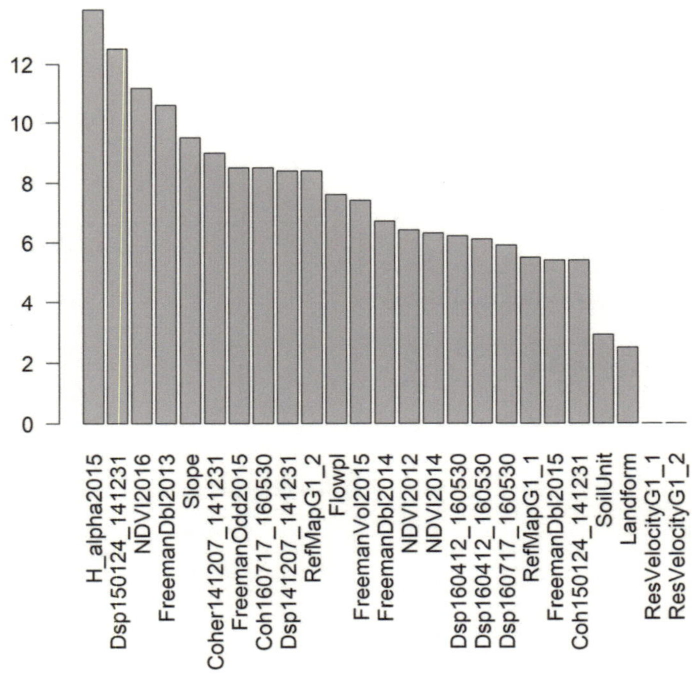

Figure 9.
Importance of the variables in decreasing order.

5. Discussion

This study confirmed that the slope angle is a key classification factor in landslide detection in a similar way reported by Donnarumffia et al. [23]. So as land use is the most influencing factor to the occurrence of landslides [24].

An analysis of the effect of DEM on InSAR processing to estimate terrain deformations showed that DEM only had a significant impact on InSAR displacement but not on InSAR coherence, such as also is highlighted in Bayer et al. [25].

Dual polarimetric SAR analysis found that VV-polarisation radar backscatter produces stronger scattering than cross-polarisation (VH) on landslide inventory in the same way as is reported by Ningthoujam et al. [26].

C/band Sentinel-1 data allowed to measure very slow ground surface displacements with mm precision by PSInSAR method. However, as is indicated in Colesanti et al. [27], it is necessary to combine data from different sources, i.e. GNSS data, to avoid misinterpretations.

Time series analysis of Landsat NDVI composites with Google Earth Engine [28], allowed to compare measurements of inter-annual NDVI. However, this research only analysed the period 2012–2016. Thus, the lacking of long-term time series of optical satellites data did not detect trends in vegetation cover changes related to landslides. For this reason, inter-annual NDVI in the period 2012–2016 only was taken as conditioning factor to develop of detection model.

Random Forest (RF) algorithm was applied to classify landslides. Conditioning factors provided by remote sensing techniques were stored as grid cells at 30 m of spatial resolution. RF model for landslide classification needed data to train the model and validate its results. The total training dataset was split with a proportion of 70% of samples used to train models and 30% for validation.

Using the test dataset, we found that the overall classification accuracy of the model was 70.8%. This meant that over 70.8% of the test dataset was correctly identified as either a landslide event or non-landslide event in the same sense as is reported in Taalab et al. [29]. The rank of variables importance, based on the relative contribution to the classification accuracy of the model, in order of importance, were: PolSAR, InSAR displacement, NDVI, backslope landform and InSAR coherence.

6. Conclusions

By using Remote Sensing techniques at the visible and microwave frequencies of EM waves this research did relate EO measurements with ground physical parameters such as scattering mechanisms, topography, land cover type and surface deformation patterns. All of the above in relationship with landslides inventory of the study area.

This research did implement unsupervised and supervised classification methods. The first to understand the pattern of LSI clustering and the second to classify the LSI with multidimensional variables derived from EO data and RS techniques.

All of the EO data collected and generated by RS techniques during this research was stored in appropriate containers of data.

This research used errors' theory, ANOVA, TUKEY and cross-validation techniques to determine the internal and external precision of the method generated for landslides detection.

Author details

Nixon Alexander Correa-Muñoz[1*†] and Carol Andrea Murillo-Feo[2†]

1 Universidad del Cauca, Popayán, Colombia

2 Universidad Nacional de Colombia, Bogotá, Colombia

*Address all correspondence to: nico@unicauca.edu.co

† These authors contributed equally.

References

[1] Martha TR., Cees J. van Westen, Norman Kerle, Victor Jetten, and K. Vinod Kumar. Landslide hazard and risk assessment using semi-automatically created landslide inventories. Geomorphology. 2013;**184**: 139-150

[2] Lee S, Min K. Statistical analysis of landslide susceptibility at Yongin. Korea. *Environmental Geology*. 2001; **40**(9):1095-1113

[3] Mărgărint MC, Grozavu A, Patriche CV. Assessing the spatial variability of weights of landslide causal factors in different regions from Romania using logistic regression. Natural Hazards and Earth System Sciences Discussions. 2013;**1**(2):1749-1774

[4] Martinis S, Clandillon S, Plank S, Twele A, Huber C, Caspard M, et al. Sadri Haouer, and Eva-Maria Fuchs. Advancing SAR and Optical Methods for Rapid Mapping. Technical Report: ASAPTERRA; January, 2017

[5] Seijmonsbergen AC, Hengl T, Anders NS. Semi-automated identification and extraction of geomorphological features using digital elevation data. In: *Developments in Earth Surface Processes*, Volume 15, Pages 297–335. Elsevier B.V. 2011

[6] Feizizadeh B, Blaschke T. Assessing uncertainties associated with digital elevation models for object based landslide Delination. In: GEOBIA 2016: Solutions and Synergies, 2016.

[7] Zhu Z. Change detection using landsat time series: A review of frequencies, preprocessing, algorithms, and applications. ISPRS Journal of Photogrammetry and Remote Sensing. 2017;**130**:370-384

[8] Simon Plank. *Rapid damage assessment by means of multi-temporal SAR-A comprehensive review and outlook to Sentinel-1,* Volume 6. 2014.

[9] Navarro J, Cuevas M, Tomás R, Barra A, Crosetto M. A toolset to detect and classify active deformation areas using interferometric SAR data. In: *Proceedings of the 5th International Conference on Geographical Information Systems Theory, Applications and Management (GISTAM 2019)*. 2019: 167-174

[10] Ferretti A, Prati C, Rocca F. Permanent scatterers in SAR interferometry. IEEE Transactions on Geoscience and Remote Sensing. 2001; **39**(1):8-20

[11] Kristina R. Czuchlewski, Jeffrey K. Weissel, and Yunjin Kim. Polarimetric synthetic aperture radar study of the Tsaoling landslide generated by the 1999 Chi-Chi earthquake, Taiwan. *Journal of Geophysical Research: Earth Surface,* 108(F1):7–17–10, 2003.

[12] Roberto Tomás and Zhenhong Li. Earth Observations for Geohazards: Present and Future Challenges. *Remote Sensing,* 9(3):194, 2017.

[13] Jacobsen K. DEMANAL program system BLUH. Technical report, Institute of Photogrammetry and GeoInformation. Hannover. 2019

[14] Kemp LD, Bonham-Carter GF, Raines GL, Looney CG. *Arc-SDM: Arcview Extension for Spatial Data Modelling Using Weights of Evidence, Logistic Regression, Fuzzy Logic and Neural Network Analysis*. Instituto de Geociências: Universidade Estadual de Campinas, São Paulo; 2001

[15] Development Core RR, Team R, Language A. Environment for statistical. Computing. 2011

[16] Yu H, Lu Z. Review on landslide susceptibility mapping using support vector machines. Catena. 2018;**165** (March):520-529

[17] Zevenbergen LW, Thorne CR. Quantitative analysis of land surface topography. Earth Surface Processes and Landforms. 1987;**12**(1):47-56

[18] Conrad O. SAGA. Program structure and current state of implementation. SAGA – Analysis and Modelling Applications. 2007;**115**:39-52

[19] Köthe R, Gehrt E, Böhner J. Automatische Reliefanalyse für geowissenschaftliche Anwendungen-derzeitiger Stand und Weiterentwicklungen des Programms SARA. Arbeitshefte Geologie. 1996;**1**:31-37

[20] Gallant J, Downling T. A multiresolution index of valley bottom flatness for mapping depositional areas. Water Resources Research. 2003;**39**(12): 1347-1360

[21] Pardo C. E., del Campo P. C., Torres C. J. Package 'FactoClass'; 2017

[22] Correa-Munoz NA, Tansey K, Murillo-Feo CA. Effect of a DEM in the estimation of coherence and unwrapped phase InSAR for landslides detection. In: *XVI Panamerican Conference on Soil Mechanics and Geotechnical Engineering.*, 2019:1693-1700.

[23] Angelo Donnarumffia, Paola Revellino, Gerardo Grelle, Francesco Maria Guadagno, and P G Revellino Grelle FM Guadagno. Slope angle as lndicator parameter of Landsli susceptibility in a geologically complex area. In C. Margottini, editor, *Landslide Science and Practice*, volume 1. Springer Berlin Heidelberg, Berlìn, 2013.

[24] Hasali Hemasinghe RSS, Rangali NLD, Samarakoon L. Landslide susceptibility mapping using logistic regression model (a case study in Badulla District, Sri Lanka). Procedia Engineering. 2018;**212**:1046-1053

[25] Benedikt Bayer, David Schmidt, and Alessandro Simoni. The influence of external digital elevation models on PS-InSAR and SBAS results: Implications for the analysis of deformation signals caused by slow moving landslides in the northern Apennines (Italy). IEEE Transactions on Geoscience and Remote Sensing, 1–14, 2017.

[26] Ramesh K. Ningthoujam, Kevin Tansey, Heiko Balzter, Keith Morrison, Sarah C.M. Johnson, France Gerard, Charles George, Geoff Burbidge, Sam Doody, Nick Veck, Gary M. Llewellyn, and Thomas Blythe. Mapping forest cover and forest cover change with airborne S-band radar. Remote Sensing, 8(7):21, 2016.

[27] Colesanti C, Wasowski J. Investigating landslides with space-borne synthetic aperture radar (SAR) interferometry. Engineering Geology. 2006;**88**(3–4):173-199

[28] Pironkova Z, Whaley R, Lan K. Time series analysis of Landsat NDVI composites with Google earth engine and R: User guide. *Science and Research Branch. Ministry of Natural Resouces and Forestry,* (December). 2018:39

[29] Taalab K, Cheng T, Yang Z. Mapping landslide susceptibility and types using Random Forest. Big Earth Data. 2018;**2**(2):1-20

Chapter 7

Comparative Evaluation of Various Statistical Models and Its Accuracy for Landslide Risk Mapping: A Case Study on Part of Himalayan Region, India

C. Prakasam, Aravinth R., Varinder S. Kanwar and B. Nagarajan

Abstract

Among other natural hazards, Landslides are the most prominent and frequently occurring natural disaster in the state of Himachal Pradesh with higher socio-economical losses. About 0.42 million sq.kms of area are prone to landslide activities in our country that is excluding the snow covered areas. The current research focuses on estimating the landslide risk zones of the Shimla Tehsil, Himachal Pradesh using various statistical models. Landslide contributing factors as such Landuse Landcover, Elevation, Slope, Lithology, Soil, Geology and Geomorphology has been used to assess the Landslide risk factors. Data obtained from LANDSAT 8 OLI sensors, SRTM DEM, Soil and Land Use Survey of India and SOI Toposheets have been used as sources. Weighted Overlay, Fuzzy logic and Analytical Hierarchical Process models will be used to categorize the Vulnerability and risk Zones of the study area. The causative factors were analyzed and processed in GIS environment. These values will be then being integrated using various studied models to produce individual landslide vulnerability and risk zones. The results reveal that most of the study area falls under Very Low risk category with a total coverage of 67.34%. Low and Moderate area covers about 23% and 9.13% of the study area. Higher risk areas only account for about 0.46%. Higher percent of the study area is mostly covered by settlements. National highways, Metal roads, Slopes and Denser settlements are located along the Moderate and low risk areas. The results retrieved from the WOM model reveals a total of 55% of the area comes under very low category. Low and Moderate category covers about 31.4% and 10.6% of the study area. High and Very High category cover a total of 1.9% together.

Keywords: weighted overlay, fuzzy logic, risk mapping, Shimla tehsil, landslide inventory

1. Introduction

"Landslides are simply defined as down slope movement of rock, debris and/ or earth under the influence of gravity. This sudden movement of material causes extensive damage to life, economy and environment" [1, 2]. Landslide occurrence in mountainous regions can be due to both natural and Man-made causes such as Roadway and Settlement construction etc. (**Figure 1**). These causes include cloud-burst, thunderstorm, construction for various activities etc. [3]. The most sensitive areas are the Himalayan belt, Western and Eastern Ghats. Among other ecosystems Mountain ecosystem is one of the most fragile ecosystem in the world, when these ecosystems are disturbed either due to natural process or Anthropogenic process or the combined effect of both results in Geohazard and environmental problems such as landslides, soil erosion, reservoir siltation and land degradation [4–7]. Among other various problems that affect hill ecosystem, landslides have observed as fast spreading epidemic due to its multivariate morphodynamic process and also due to improper interaction of human being on nature, especially terrain ecosystem [8–10]. Not only In India but countries located along the borders of Himalayan region experience frequent landslide occurrences. These areas include the North Western and North Eastern Himalayas and the Western & Eastern Ghats [3]. Statistics on world landslides and its impacts is given in **Table 1**.

Landslides of Various types, Pose severe hazards in the Himalayan mountains. Some of the worst disasters in the world have been caused by landslides [11–14]. Every year the damage cause by caused landslide amounts 1 billion$ says a us study and an average of 200 deaths in a year which is 30% of the such losses around the world [15, 16]. Map derived from the world bank data [17] indicates that most of the landslide hotspots are located the along the mountainous region and active tectonics region (**Figure 2**) [18].

Surya Prakash [19] has compiled a list of landslides that have socio- economic impacts in India (**Table 2**). In his research he reported that the Western and North Western Himalayas account for about 51% high prone landslides.

Over the years' various scientist and researchers have addressed the landslide problems using various qualitative, Quantitative, Statistical and Numerical

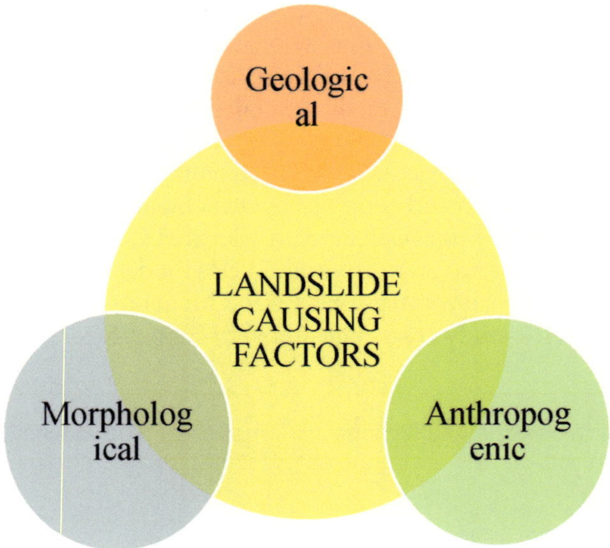

Figure 1.
Landslide causative factors.

Continents	Events	Killed	Injured	Homeless	Affected	Total affected
Africa	23	745	56	7936	13,748	21,740
Avg. Per event		32	2	345	568	945
America	145	20,684	4809	1,86,752	44,85,037	46,76,598
Avg. Per event		143	33	1288	30,931	32,252
Asia	255	18,299	3776	38,25,311	16,47,683	54,76,770
Avg. Per event		72	15	15,001	6462	21,478
Europe	72	16,758	523	8625	39,376	48,524
Avg. Per event		23	7	120	547	674
Oceania	16	542	52	18,000	2963	21,015
Avg. Per event		34	3	1125	185	1313

Table 1.
World statistics on landslides (1900–2010).

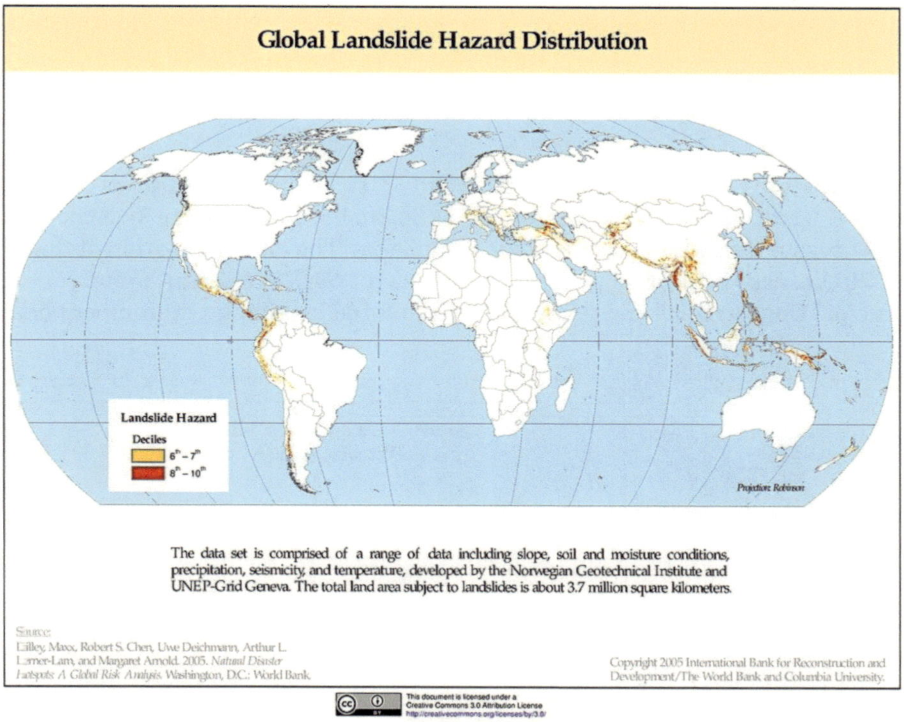

Figure 2.
Global landslide hotspots (2017), source (World Bank).

methods. Every methodology used has its own merits and demerits. The type of landslide involved depends upon varying condition such as slope, geology, geomorphology, nature of landslide, type of causative factor etc [16, 20, 21]. Based on the literature it can be concluded that no methodology is global for landslide studies. The accuracy of estimating landslide risk zones various for each method [22–28]. It is imperative reliable methodology should be used for the analysis of regional scale landslide risk mapping. Addressing the nature and causative factor of individual landslide is as important as preparing landslide risk maps. Most of the literature address the mapping risk zones or slope stability analysis, it is imperative that all

Sl.no	Year	No. of socio economically significant events	Persons killed	No. of fatal events
1	2018–2019	1	0	0
2	2007–2017	893 (Nasa Catalog)	6614	893
3	2011	26	74	19
4	2010	85	368	53
5	2009	47	270	46
6	2008	36	220	30
7	2007	54	409	39
8	1800–2007	123	2630	61

Table 2.
Socio-economic significant landslides (1800–2011).

these problems should be addressed together. The current research is focused on analyzing the accuracy of weighted Overlay model and Fuzzy logic model to estimate the landslide risk mapping along the Rampur tehsil, Himachal Pradesh, India.

2. Study area

The study area extends from "76°58′19" to 77°19′21″ longitude and 30°59′3″ to 31°14′10″ h latitude" with a total area of 368 Sq.km hectares (**Figure 3**). According to 2011 census the Shimla has a total of 576 villages. The total population of Shimla as of 2011 census is 1,71,640 people among which 1,69,578 reside in "Shimla Municipal Corporation" and the rest in Shimla Rural and Jutogh cantonment board.

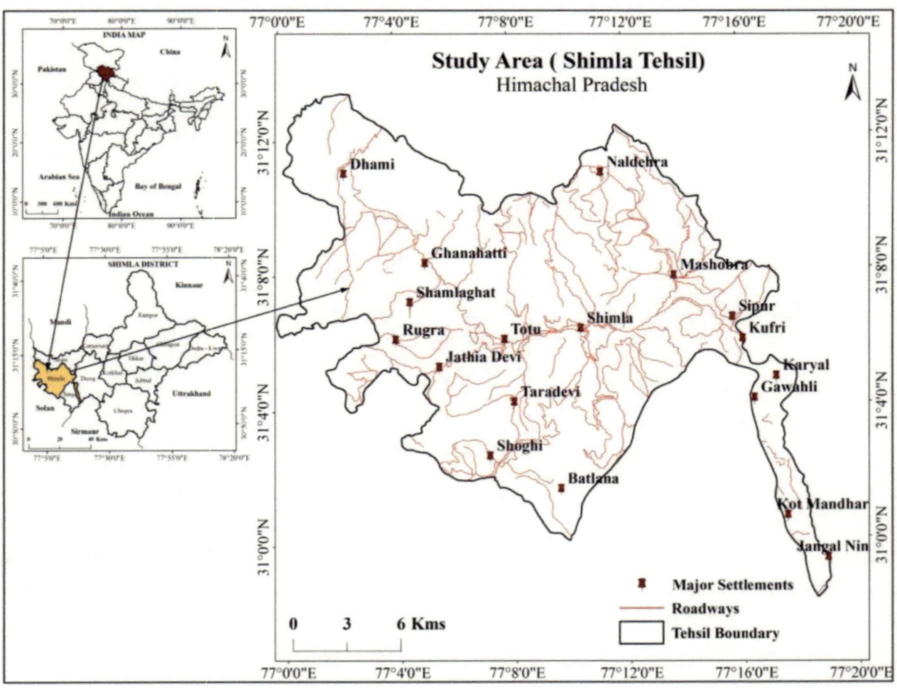

Figure 3.
Study area.

Sutlej, Giri and Pabbar are the three major rivers that drain through Shimla. The economic activities are majorly dependent upon agriculture, horticulture and tourist activities in these areas. Cereals, Off season vegetables and stone fruits are most suitable to grow in the high altitude areas. Most of the Agriculture is rainfall dependant. The soil varies from Sandy loam to Loamy skeletal in the valley and mountain regions. Geologically. Lithology was interpreted from the maps retrieved form the Soil and Landuse survey of India (SLUSI). Major rock types present in these area are Granite, Phyllite, Dolomite, Limestone and Shale. Geomorphologically the area is mostly Undifferentiated hill side and mountain side slope. An average of 999.4 mm of rainfall is recorded where most of the rainfall is received during monsoon period. "The temperature can go as low as 0°C during winter times and as much as 40°C during summer times".

3. Dataset and research method

The base map of the study area was digitized from Survey of India Toposheets. One cloud free satellite data LANDSAT 8OLI (26/01/2020) was downloaded from the earth explorer website. Soil data covering the study area was received from "Soil and Landuse Survey of India (SLUSI)". In addition, a 30 mts ASTERGDEM data was downloaded from USGS website for topographical analysis. Rainfall data has been acquired from Indian Meteorological Department, Shimla. The types of data used is given in (**Table 3**).

Weighted Overlay and Fuzzy logic models are the two type statistical methods used in the research. In the recent years many researchers and scientist have used the methodology to derive landslide risk mapping with higher accuracies [29–36]. "Barrile Vincenzo et.al, 2016 used Fuzzy logic method for mapping landslide susceptibilities. The province of Reggio Calabria, Italy chosen as study area. Parameters such as Elevation Slope, Lithology, Rainfall and Landuse were assigned values and processed in GIS environment. The output subdivides into five categories ranging from Very low to Very high. The results indicate that 22%, 36% and 20% of the area comes under Very high, High and Moderate risk zones". "Leonardi Geovani et.al, 2016 used a Fuzzy approach to analyze landslide susceptibility for Reggio Province, Calabria, Italy. Rainfall, Elevation, Slope, Landuse and Lithology were used as landslide influencing parameters. The output signifies that 22% and 36% of the area

Sl. No	Data	Source	Date	Resolution
1	Toposheets	SOI	1987	1:50,000
2	Rainfall	IMD	2000–2017	—
3	Soil	"Soil and land use survey of India"	—	1:50,000
4	Geology	"Soil and land use survey of India"		1:50,000
5	Geomorphology	"Soil and land use survey of India"		1:50,000
6	LULC	Landsat 8 OLI USGS	26/01/2020	30 mts
7	ASTERGDEM	USGS	2009	30 mts
8	Landslide Inventory	Google Earth	2017	0.4 mts

Table 3.
Data used.

comes under high and very high risk areas. The results were validated with accuracy of 80% with the data". The fuzzy logic method uses a value of 0 to 1 to evaluate the relation between landslide occurrences with it respective causative factors. Then the causative factors are analyzed and integrated in the GIS environment to create landslide risk maps and landslide inventory data collected from the field is used to establish the degree of association with each causative factors. "Weighted Overlay Model (WOM)" uses numerical based rating method to classify the parameters ranging from very low to very high based on its degree of importance for landslide initiation and each sub factor is classified into sub categories at a scale of 1 to 5 where 1 indicating the very low risk and 5 indicates very high risk.

4. Result and discussion

4.1 Landuse and landcover

Landuse and Landcover is one of the important causative factor for landslide risk and initiation. Urbanization along hilly areas and unstable constructions lead to slope instability causing slope failure. The LULC has interpreted from LANDSAT OLI imagery for the year 2020. Supervised classification method has been use to map Land cover of the study area. Shimla Tehsil has been classified into four classes namely Forest, Agriculture, Slopes and Settlement (**Table 4**). The classification was based NRSC Level I classification system of Landuse.

Among the various land covers forest is comprised of 63.4% of the total area. Agriculture and barren land together makes 26.7% of the study area. Settlement account for only about 9.7% of the study area (**Figure 4**). The occurrences of mass movements of landslides are minimal along forest due to its soil binding capacity. Tree root bind the soil to the ground avoiding soil erosion due to torrential and monsoon rainfall. In places such as slopes and settlements soils are exposed without vegetation cover and hence prone to a large No. of landslides during monsoon season in the stud area.

4.2 Soil

Soil plays an active role a landslide control factor especially in rugged terrains such as Himalayas. In the study area the soil class area differentiated into coarse loamy, fine loamy and loamy skeletal (**Table 5**). Fine loamy soils in these areas make up about 95.6% of the study area. The rest of the soils coarse loamy and loamy skeletal makes up about 4.2% of the area. Fine soil loamy soil has as clay content of between 18 to 35% and the reminder covered by sand and silt (**Figure 5**). Fine loamy

Sl. no	Class	Area (hectares)	Percent (%)	Weighted overlay model	Fuzzy logic
1	Agriculture	5225.31	14.19%	2	0.3
2	Forest	23,354.53	63.42%	1	0.1
3	Build-up	3605.37	9.79%	4	0.7
4	Slope	4639.24	12.60%	5	0.8
	Total	36,824	100.00%		

Table 4.
LULC with WOM and fuzzy overlay values.

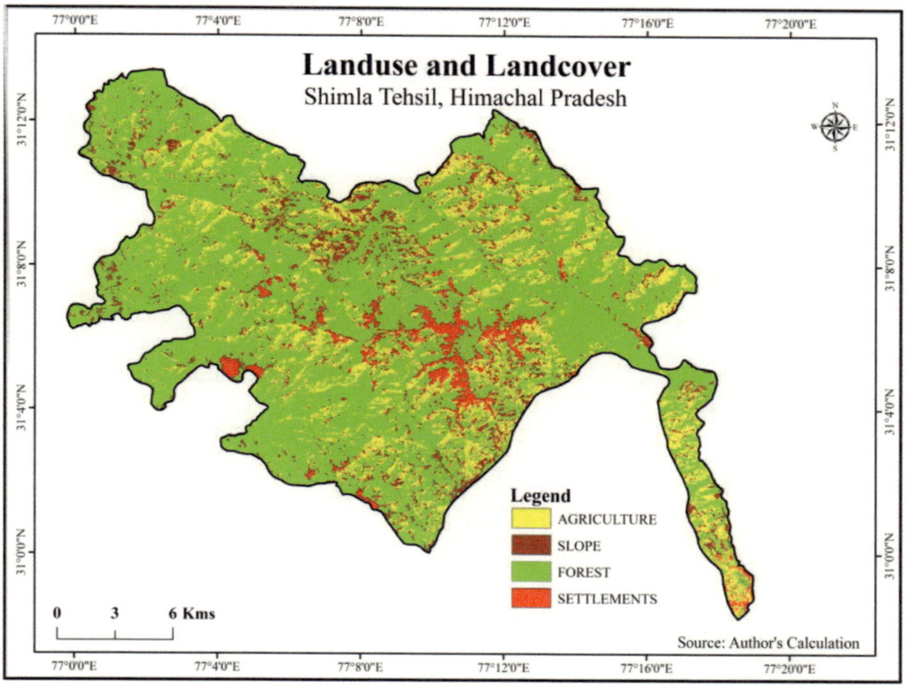

Figure 4.
Landuse and Landcover.

Sl. no	Soil class	Area (sq.km)	Percent coverage (%)	Weighted overlay model	Fuzzy logic
1	Coarse Loamy	6	1.63%	4	0.6
2	Fine Loamy	352	95.65%	3	0.4
3	Habitation	5	1.36%	4	0.8
4	Loamy Skeletal	5	1.36%	3	0.5
		368	100.00%		

Table 5.
Soil with WOM and fuzzy overlay values.

soils are moderately prone to landslides due to their high clay content. These types of soils are prone to mass movements when the water stress in the soil particle exceeds the effective stress of the soil. Coarse loamy and loamy skeletal soil have less clay content between 10 to 18% which are considerably more prone to landslides than fine loamy soil. These soil particles without any vegetation cover are more susceptible to mass movements when the deformation rate in the ground is high.

4.3 Geomorphology

The geomorphology has been differentiated into Habitation, Undifferentiated Hillside and Mountainside slopes (**Table 6**). Undifferentiated mountainside slopes cover about 97.83% and the hillside slopes and habitation covers only minor quantities about 0.5% and 1.6% in the study area (**Figure 6**). Geomorphologically Shimla Tehsil is moderately prone to landslides while the settlement areas are highly prone to mass movement.

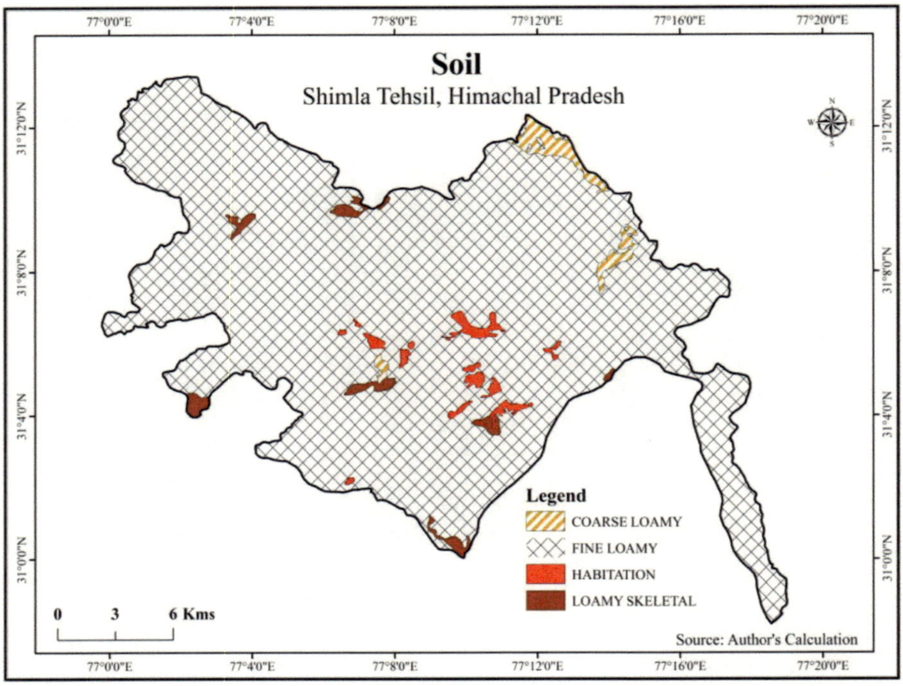

Figure 5.
Soil.

Sl.no	Soil class	Area (sq.km)	Percent coverage (%)	Weighted overlay model	Fuzzy logic
1	Habitation	6	1.63%	4	0.8
2	Undifferentiated Hill Side Slope	2	0.54%	3	0.5
3	Undifferentiated Mountain Side Slope	360	97.83%	3	0.5
		368	100.00%		

Table 6.
Geomorphology with WOM and fuzzy overlay values.

4.4 Geology

Geology plays a key role in groundwater recharge as the types rocks present in an area could hugely affect the amount of water entering into the groundwater table. The study area is covered three major classes namely Schist, Slate and Habitation (**Figure 7**) in which mainly dominated by two types of rock formation namely Slate and Schist that comprises 95.9% and 2.1% (**Table 7**) of the study area. These rocks vary from Moderate to strong in nature in the GSI index. Waters can percolate through the cracks within the rocks or even between them. Fractures and joints formed along the rock surface act as perfect carriers for Rainwater into the groundwater table. Habitation accounts for only 1.90% and these areas mostly comprised of settlements they are placed in the high risk factor for mass movements initiation.

4.5 DEM

Elevation is a secondary factor often used in landslide risk mapping. The risk mapping in DEM model depends upon the no of landslides occurring within a particular elevation height. In the current study area most of the landslide occurs between 1400 to 2100 mts (**Table 8**) above the mean sea level compared to other elevation heights. Hence these particular elevations are assigned higher risk values (**Figure 8**).

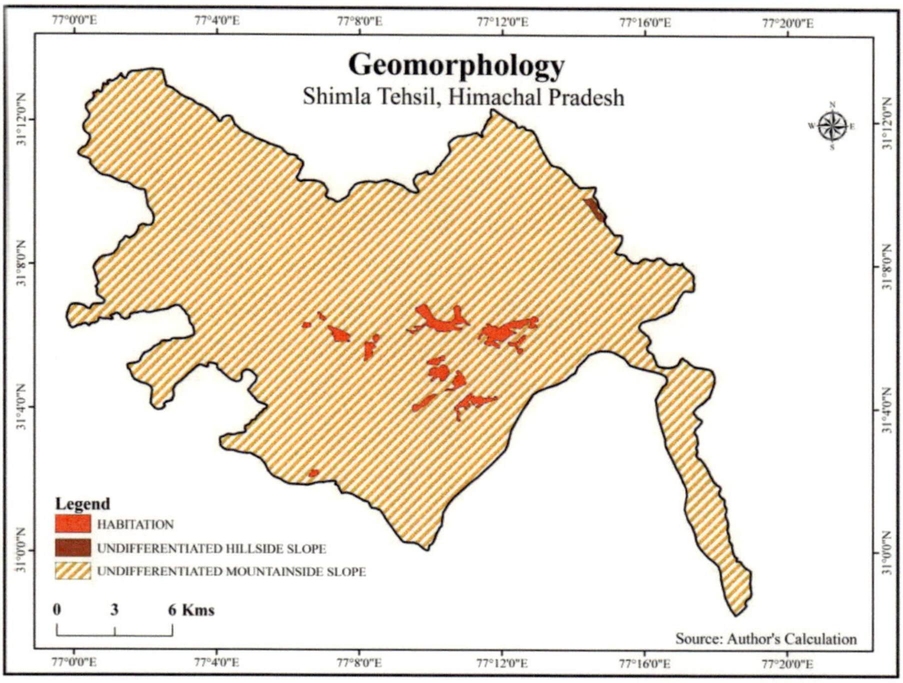

Figure 6.
Geomorphology.

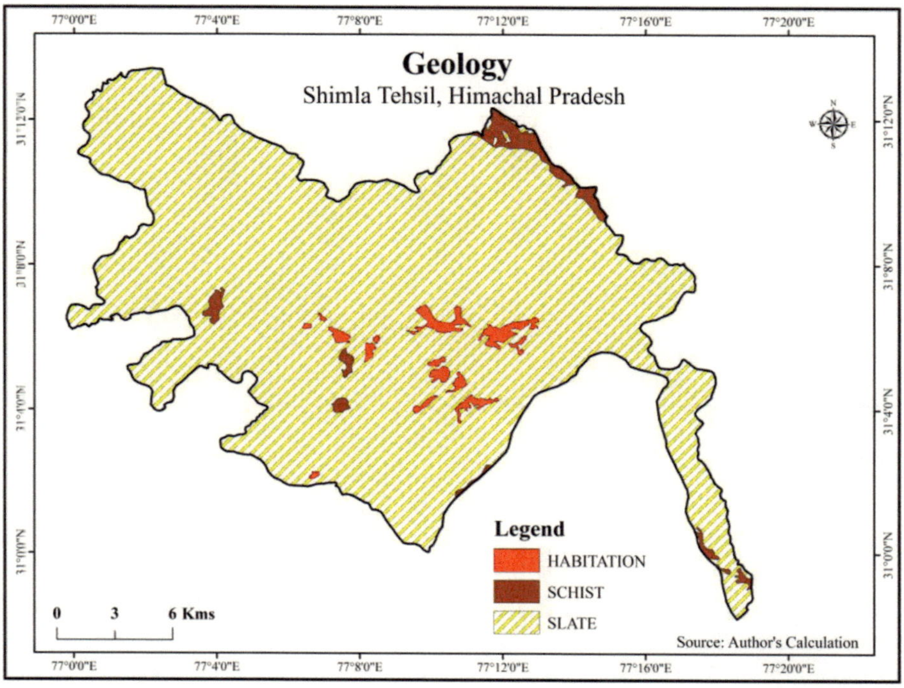

Figure 7.
Geology.

Sl. no	Soil class	Area (sq.km)	Percent coverage (%)	Weighted overlay model	Fuzzy logic
1	Habitation	7	1.90%	4	0.8
2	Schist	8	2.17%	3	0.6
3	Slate	353	95.92%	3	0.4
		368	100.00%		

Table 7.
Geology with WOM and fuzzy overlay values.

Sl. no	Elevation (mts)	Weighted overlay model	Fuzzy logic
1	820 to 1300	1	0.3
2	1301 to 1600	3	0.6
3	1601 to 1800	4	0.8
4	1801 to 2100	4	0.7
5	2101 to 2200	2	0.3

Table 8.
DEM with WOM and fuzzy overlay values.

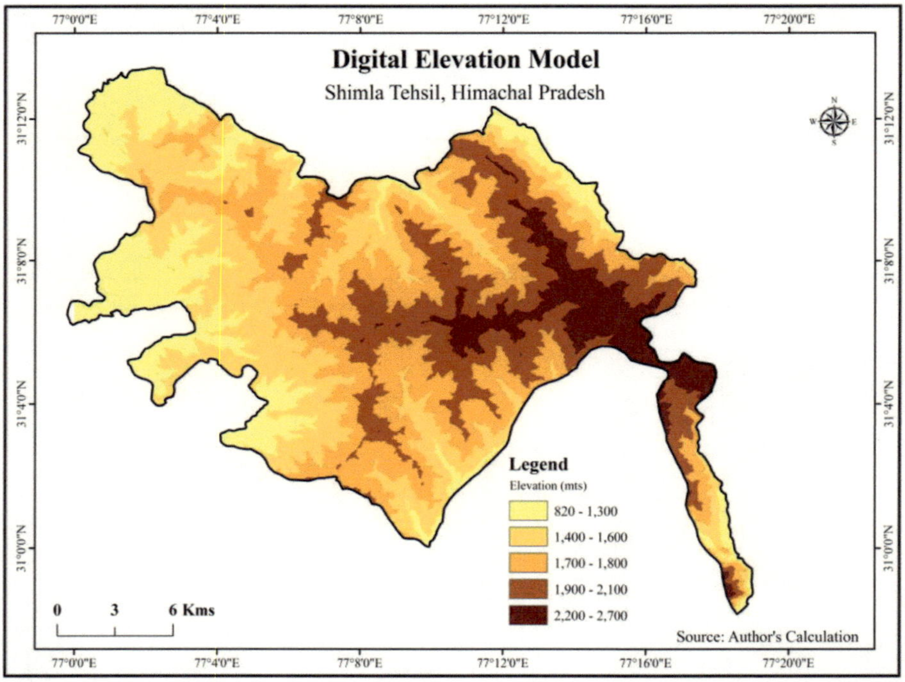

Figure 8.
Digital elevation model.

4.6 Slope

Slope aspect plays a crucial in highly dissected mountainous regions for landslide movements. The steeper the angle of the slopes the higher the possibility of the mass movements. In the research SRTM DEM data has been used for deriving slope parameters (**Figure 9**). The slopes has been classified into five ranging from very gentle to very steep (**Table 9**) in nature.

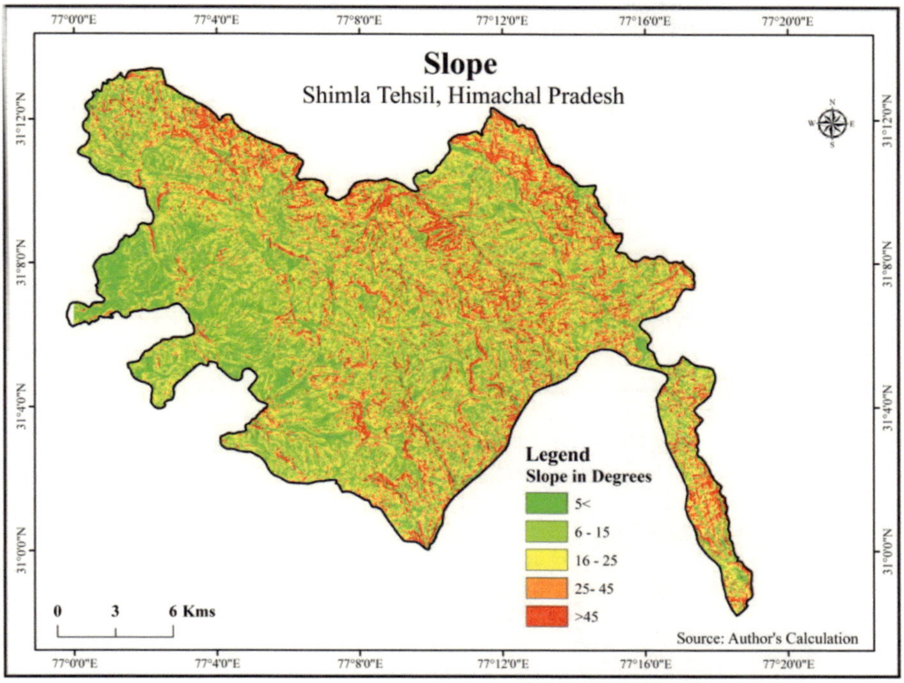

Figure 9.
Slope.

Sl. no	Slope (degrees)	Angle	Weighted overlay model	Fuzzy logic
1	< 5	Very gentle	1	0.1
2	6 to 15	Moderately gentle	2	0.3
3	16 to 25	Moderately steep	3	0.5
4	26 to 45	Very steep	4	0.8
5	45	Extremely steep	5	0.9

Table 9.
Slope with WOM and fuzzy overlay values.

4.7 Landslide risk mapping

4.7.1 Fuzzy logic model

For the current study two statistical models were employed "Fuzzy logic and Weighted Overlay model (WOM)" for landslide risk mapping. Factor including LULC, Geology, Geomorphology, Soil, slope (**Table 10**) were used as factoring parameters for risk mapping. Each causative factors were assigned a value of 0 to 1 based on it degree of association between causative factors. The factors are then processed in the GIS environment to derive fuzzy logic based landslide risk mapping.

Based on the results it can be concluded that most of the study area falls under very low risk with a total coverage of 67.34%. Low and Moderate area covers about 23% and 9.13% of the study area. Higher risk areas only account for about 0.46%. Higher percent of the study area is mostly covered by settlements. National highways, Metal roads, Slopes and Denser settlements are located along the Moderate and low risk areas (**Figure 10**).

Sl. no	Landslide risk class	Area (sq.km)	Percent coverage (%)
1	Very low	247.8	67.34%
2	Low	84.9	23.07%
3	Moderate	33.6	9.13%
4	High	1.7	0.46%
		368	100.00%

Table 10.
Fuzzy logic landslide risk categories.

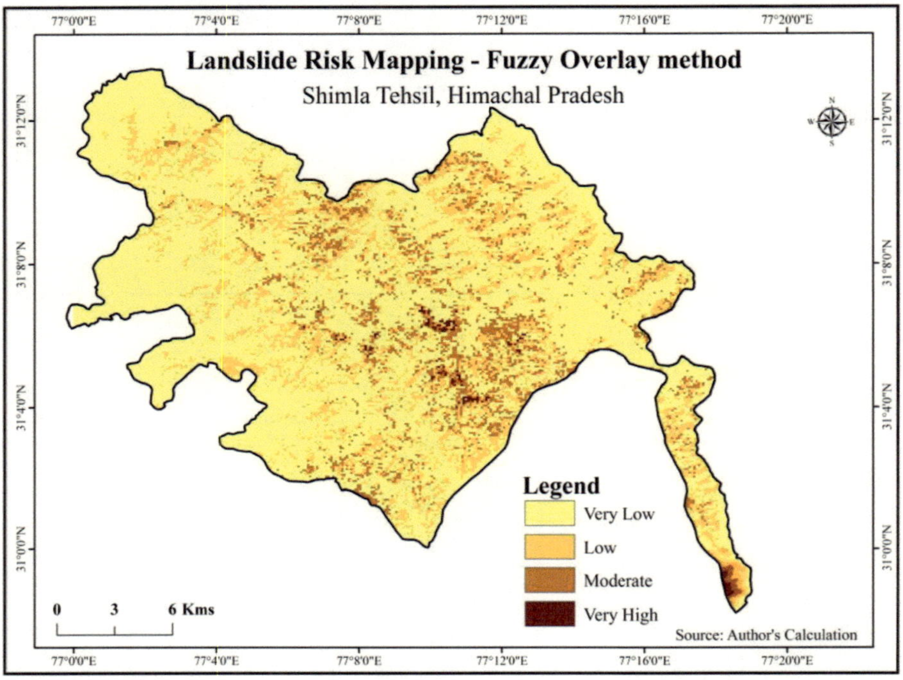

Figure 10.
Fuzzy logic based landslide risk assessment.

4.7.2 Weighted overlay model

Weighted Overlay Model (WOM) was used as a second statistical method for Landslide Risk mapping. Weighted overlay model uses the ranking method to classify each causative factors based on its degree of importance for landslide initiation and each sub factor is classified into sub categories at a scale of 1 to 5 where 1 indicating the very low risk and 5 indicates very high risk. Six causative factors namely LULC, Geology, Geomorphology, Soil (**Table 11**) etc. was used. The factors are then processed in the GIS environment to derive fuzzy logic based landslide risk mapping (**Figure 11**).

The results retrieved from the WOM model reveals 55% of the Tehsil comes under very low category. Low and Moderate category covers about 31.4% and 10.6% of the study area. High and Very High category cover a total of 1.9% together. Most of the low category indicators are located along the Forest and Agricultural areas that include plantations. Slope and settlements covers a major part of Moderate to Very High vulnerable areas.

Sl. no	Landslide risk class	Area (sq.km)	Percent coverage (%)
1	Very low	205.726	55.90%
2	Low	115.829	31.47%
3	Moderate	39.028	10.61%
4	High	6.547	1.78%
5	Very high	0.876	0.24%
		368	100.00%

Table 11.
Weighted overlay based landslide risk categories.

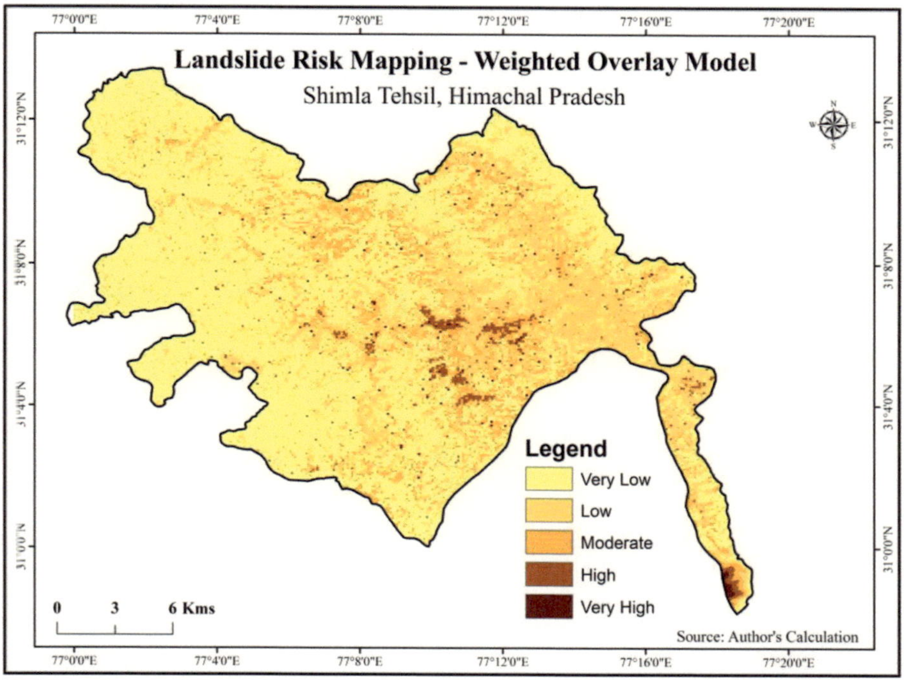

Figure 11.
Weighted overlay model based landslide risk assessment.

5. Conclusion

The research reveals the landslide risk mapping of the study area through Weighted Overlay and Fuzzy logic models. Fuzzy model classified a total of 9.1% and 0.4% of the area under moderate and high risk categories whereas Weighted Overlay model classified a total of 10% and 2% area under moderate and high to very high risk categories. Both the statistical model covers Forest and Agricultural areas under Very Low to Low Risk factor. Areas located along barren lands, Settlements and Roadways are classified under moderate to Very High risk areas for mass movements.

Acknowledgements

"The research work done is a part of NRDMS-DST funded research project. We would like to express our sincerest gratitude to NRDMS-DST, GOI, New Delhi, India for funding this research project".

"We would like to thank CSIR, New Delhi, GOI for the SRF – Direct, scholarship for pursing Research Work".

Author details

C. Prakasam[1]*, Aravinth R.[1], Varinder S. Kanwar[1] and B. Nagarajan[2]

1 Department of Civil Engineering, Chitkara University, Himachal Pradesh, India

2 National Centre for Geodesy, Indian Institute of Technology Kanpur, Uttar Pradesh, India

*Address all correspondence to: cprakasam@gmail.com

References

[1] Varnes DJ. Landslide types and processes. Landslides and engineering practice. 1958;24:20-47.

[2] Varnes DJ. Slope movement types and processes. Special report. 1978;176:11-3

[3] Introduction to Indian Landslides and Statistics. Available from: https://www.gsi.gov.in/webcenter/portal/OCBIS/pageGeoInfo/pageLANDSLIDEHAZRD?_afrLoop=1714095558132139&_adf.ctrl-state=z4iu12hiz_1#!%40%40%3F_afrLoop%3D1714095558132139%26_adf.ctrl-state%3Dz4iu12hiz_5

[4] Mondal S, Mandal S. RS & GIS-based landslide susceptibility mapping of the Balason River basin, Darjeeling Himalaya, using logistic regression (LR) model. Georisk [Internet]. 2018;12(1):29-44. Available from: doi:10.1080/17499518.2017.1347949

[5] Kumar A, Asthana AKL, Priyanka RS, Jayangondaperumal R, Gupta AK, Bhakuni SS. Assessment of landslide hazards induced by extreme rainfall event in Jammu and Kashmir Himalaya, northwest India. Geomorphology [Internet]. 2017;284:72-87. Available from: doi:10.1016/j.geomorph.2017.01.003

[6] Ashutosh PKS, Panthee KS. Rockfall analysis along transportation corridors in high hill slopes. Environ Earth Sci. 2016;1-11.

[7] Society IM. Landslides and Erosion in the Catchment of the Gaula River, Kumaun Lesser Himalaya, India Author (s): S. K. Bartarya and K. S. Valdiya Published by: International Mountain Society Stable URL : http://www.jstor.org/stable/3673588REFERENCES Linked references are available on JSTOR for this article : You may need to log in to JSTOR to access the linked references. LANDSLIDES AND EROSION IN THE CATCHMENT OF THE GAULA RIVER, KUMAUN LESSER HIMALAYA, INDIA. 2016;9(4):405-19.

[8] Sarkar S, Roy AK, Raha P. Deterministic approach for susceptibility assessment of shallow debris slide in the Darjeeling Himalayas, India. Catena [Internet]. 2016;142:36-46. Available from: doi:10.1016/j.catena.2016.02.009

[9] Metha RL, Koli SR, Koli VR. Landslide Hazard Zonation Using Remote Sensing and GIS Technology : A Case Study of Landslide Prone Area near Mahabaleshwar, Maharashtra,. 2015;3(4):6-16.

[10] Sharma G, Sanjeevi S. Landslide Hazard Zonation using Remote Sensing, Ground Penetrating Radar surveys and Geographical Information System in Katteri. Int J Curr Eng Technol. 2015;5(2):1160-9.

[11] Ghosh S, Carranza EJM, van Westen CJ, Jetten VG, Bhattacharya DN. Selecting and weighting spatial predictors for empirical modeling of landslide susceptibility in the darjeeling himalayas (india). Geomorphology [Internet]. 2011;131(1-2):35-56. Available from: doi:10.1016/j.geomorph.2011.04.019

[12] Kanungo DP, Sarkar S, Sharma S. Combining neural network with fuzzy, certainty factor and likelihood ratio concepts for spatial prediction of landslides. Nat Hazards. 2011;59(3):1491-512.

[13] Sharma LP, Patel N, Ghose MK, Debnath P. Landslide vulnerability assessment and zonation through ranking of causative parameters based on landslide density-derived statistical indicators. Geocarto Int. 2011;26(6):491-504.

[14] Martha TR, Kerle N, van Westen CJ, Jetten V, Vinod Kumar K. Object-oriented analysis of multi-temporal panchromatic images for creation of historical landslide inventories. ISPRS J Photogramm Remote Sens [Internet]. 2012;67(1):105-19. Available from: doi:10.1016/j.isprsjprs.2011.11.004

[15] Bilham R, Bali BS. A ninth century earthquake-induced landslide and flood in the Kashmir Valley, and earthquake damage to Kashmir's Medieval temples. Bull Earthq Eng. 2014;12(1):79-109.

[16] Pareek N, Pal S, Sharma ML, Arora MK. Study of effect of seismic displacements on landslide susceptibility zonation (LSZ) in Garhwal Himalayan region of India using GIS and remote sensing techniques. Comput Geosci [Internet]. 2013;61:50-63. Available from: doi:10.1016/j.cageo.2013.07.018

[17] Global landslide hotspots (2017). Available from: https://datacatalog.worldbank.org/dataset/global-landslide-hazard-map

[18] Naithani AK. The Himalayan landslides. *Employment News*. 1999;*23*(47):20-26

[19] Prakash S. Historical Records of Socio-Economically Significant Landslides in India. *Journal of South Asia Disaster Studies*. 2011;**4**(2):177-204

[20] Martha TR, van Westen CJ, Kerle N, Jetten V, Vinod Kumar K. Landslide hazard and risk assessment using semi-automatically created landslide inventories. Geomorphology [Internet]. 2013;184:139-50. Available from: doi:10.1016/j.geomorph.2012.12.001

[21] Singh R, Umrao RK, Singh TN. Probabilistic analysis of slope in Amiyan landslide area, Uttarakhand. Geomatics, Nat Hazards Risk. 2013;4(1):13-29.

[22] Chauhan S, Sharma M, Arora MK, Gupta NK. Landslide susceptibility zonation through ratings derived from artificial neural network. Int J Appl Earth Obs Geoinf [Internet]. 2010;12(5):340-50. Available from: doi:10.1016/j.jag.2010.04.006

[23] Martha TR, Kerle N, Jetten V, van Westen CJ, Kumar KV. Characterising spectral, spatial and morphometric properties of landslides for semi-automatic detection using object-oriented methods. Geomorphology [Internet]. 2010;116(1-2):24-36. Available from: doi:10.1016/j.geomorph.2009.10.004

[24] Das I, Sahoo S, van Westen C, Stein A, Hack R. Landslide susceptibility assessment using logistic regression and its comparison with a rock mass classification system, along a road section in the northern Himalayas (India). Geomorphology [Internet]. 2010;114(4):627-37. Available from: doi:10.1016/j.geomorph.2009.09.023

[25] Mathew J, Jha VK, Rawat GS. Weights of evidence modelling for landslide hazard zonation mapping in part of Bhagirathi valley, Uttarakhand. Curr Sci. 2007;92(5):628-38.

[26] Ray PKC, Dimri S, Lakhera RC, Sati S. Fuzzy-based method for landslide hazard assessment in active seismic zone of Himalaya. Landslides. 2007;4(2):101-11.

[27] Gupta RP, Kanungo DP, Arora MK, Sarkar S. Approaches for comparative evaluation of raster GIS-based landslide susceptibility zonation maps. Int J Appl Earth Obs Geoinf. 2008;10(3):330-41.

[28] Kanungo DP, Arora MK, Sarkar S, Gupta RP. A comparative study of conventional, ANN black box, fuzzy and combined neural and fuzzy weighting procedures for landslide susceptibility zonation in Darjeeling Himalayas. Eng Geol. 2006;85(3-4):347-66.

[29] Feizizadeh B, Shadman Roodposhti M, Jankowski P, Blaschke T. A GIS-based extended fuzzy multi-criteria evaluation for landslide susceptibility mapping. Comput Geosci [Internet]. 2014;73:208-21. Available from: doi:10.1016/j.cageo.2014.08.001

[30] Kayastha P, Dhital MR, De Smedt F. Application of the analytical hierarchy process (AHP) for landslide susceptibility mapping: A case study from the Tinau watershed, west Nepal. Comput Geosci [Internet]. 2013;52:398-408. Available from: doi:10.1016/j.cageo.2012.11.003

[31] Alimohammadlou Y, Najafi A, Yalcin A. Landslide process and impacts: A proposed classification method. Catena. 2013;104:219-32.

[32] Raghuvanshi TK, Negassa L, Kala PM. GIS based Grid overlay method versus modeling approach - A comparative study for landslide hazard zonation (LHZ) in Meta Robi District of West Showa Zone in Ethiopia. Egypt J Remote Sens Sp Sci [Internet]. 2015;18(2):235-50. Available from: doi:10.1016/j.ejrs.2015.08.001

[33] Guzzetti F, Cesare A, Cardinali M, Fiorucci F, Santangelo M, Chang K. Earth-Science Reviews Landslide inventory maps : New tools for an old problem. Earth Sci Rev [Internet]. 2012;112(1-2):42-66. Available from: doi:10.1016/j.earscirev.2012.02.001

[34] Feizizadeh B, Blaschke T, Tiede D, Moghaddam MHR. Evaluating fuzzy operators of an object-based image analysis for detecting landslides and their changes. Geomorphology [Internet]. 2017;293(June):240-54. Available from: doi:10.1016/j.geomorph.2017.06.002

[35] Catani F, Tofani V, Lagomarsino D. Spatial patterns of landslide dimension: A tool for magnitude mapping. Geomorphology [Internet]. 2016;273:361-73. Available from: doi:10.1016/j.geomorph.2016.08.032

[36] Khattak GA, Owen LA, Kamp U, Harp EL. Evolution of earthquake-triggered landslides in the Kashmir Himalaya, northern Pakistan. Geomorphology [Internet]. 2010;115(1-2):102-8. Available from: doi:10.1016/j.geomorph.2009.09.035

Chapter 8

Integrated Analysis Method for Stability Analysis and Maintenance of Cut-Slope in Urban

Mincheol Park, Heuisoo Han and Yoonhwa Jin

Abstract

In the process of constructing roads for the development of the city, cut-slopes are made by excavating mountains. However, these cut-slopes are degraded in strength by time-deterioration phenomenon, and progressive slope failure is caused. This study developed an integrated analysis method for stability analysis and maintenance of cut-slopes in urban. The slope stability analysis was performed using the finite element model, and the progressive slope failure by time-dependent deterioration was quantified by using the strength parameters of soil applying the strength reduction factor (SRF). The displacements until the slope failure by slope stability analysis were quantified by cumulative displacement curve, velocity curve, and inverse velocity curve and, applied to the slope maintenance method. The inverse-velocity curve applied to the prediction of the time of slope failure was regressed to the 1st linear equation in the brittle material and the 3rd polynomial equation in the ductile material. This is consistent with the proposed formula of Fukuzono and also shows similar behavior to the failure case in literature. In the future, integrated analysis method should be improved through additional research. And it should be applied to cut-slope to prevent disasters.

Keywords: cut-slope, slope maintenance, slope stability analysis, progressive slope failure, time-dependent deterioration of slope

1. Introduction

Recently, there have been many natural disasters due to climate change. In particular, slope failure in urban areas has caused loss of lives and of property [1]. The causes of slope failures around the world are intense rainfall, rapid snowmelt, water level changes in rivers or lakes at the foot of slopes, volcanic eruptions, and earthquakes [2]. In the type of slope failure, Soil slope mainly was occurred a deep circular failure and a shallow plane failure. The failure of the rock slope is caused by activity, overturning, and rockfall. In addition, there is debris flow and creep in soft ground [3].

The types of slope failure are determined by the composting materials and exterior conditions of slope. The shear strength of slope soil is decreased as time goes by, especially time-dependent deterioration of cut-slope would greatly happen after excavation. These cut-slopes are degraded in strength by time-deterioration phenomenon, and progressive slope failure is caused.

A time-dependent deterioration of soil owing to external environmental change causes progressive slope failure and the traditional analysis of limit equilibrium stability has limitations for an appropriate analysis of the stability [4]. The analysis of progressive slope failure requires finite element analysis that is capable of analyzing the creation and progression of shearing zone [5–7]. The behavior of progressive slope failure can be evaluated through the finite element analysis. Besides, for the analysis of the slope stability, the calculation of safety factor is needed. Zeinkiewicz, et al. [8] have proposed the strength reduction method to calculate the safety factor through finite element analysis and the proposition was followed by many subsequent studies conducted by many researchers [9–11]. Recently, there has been many studies unsaturated slope stability analysis induced by rainfall infiltration [12–16]. In general, rainfall-induced slope failures are caused by increased pore pressure and seepage force during periods of intense rainfall [17, 18]. The factor of safety on the slope is calculated by the equilibrium equation of the force of the failure surface. The pore pressure acting as an active force on the failure surface is increased by seepage of rainfall, and the slope is collapsed when it is larger than the resistance force.

The slope behaviors and exterior environment should be measured continuously to maintain slope. The slope behavior is measured by inclinometer, electro-optical wave distance measuring instrument, groundwater level meter etc., also, the exterior environments, rainfall and temperature are measured by the weather station. The various management criteria of the sensors are developed based on mathematical or statistical methods; the slope reinforcement be done if the management criteria [19]. However, the developments of perfect safety factors and management criteria are very difficult from the analysis of measured data in slope.

In Korea, a lot of researches were done for cut-slope management near roadway [20–22]. The displacements of cut-slope were measured by tension wire, and the data were analyzed by the statistical process control (SPC) method. However, it confirm only the abnormal behaviors of slopes, the factor of safety (FOS) of slope cannot be calculated, therefore, the application of slope reinforcement methods are determined according to extra slope stability analysis.

Also, many studies had also been carried out to predict the time of the failure of slope through displacement velocity [1, 23–28]. The researchers have used the inverse-velocity obtained from measurements to predict the time of the slope failure and verified respective applications of the inverse-velocity through actual cases of the slope failure and experiments. However, in cases of slopes, the prediction of the time of the slope failure by using the inverse-velocity curve derived from such measurement is quite difficult because the behavioral aspects of such slope are diverse and the failure surface has to be assumed. So far, the slope stability analysis methods only estimate the FOS of slope and, maintenance methods based on the measured data do not provide clear management criteria. For these reasons, efficient maintenance and prevention of slope failure in urban areas have not been achieved. To solve this problem, the integrated analysis methods of slope stability analysis and maintenance for progressive slope failure due to time-dependent deterioration should be developed.

This study developed an integrated analysis method for stability analysis and maintenance of cut-slopes in urban. The integrated analysis method for this research treated cut-slope and the failure-inducing factor was considered the shear strength decrease (SRF) by time-dependent deterioration was considered as inducing factor of slope failure. The strength variation of soil slope happens continuously from the variation of weather conditions and infiltration of rainfall. Also, measuring the slope displacements is easier then sensing the capillary force and water content of slope. Therefore, for this research, the slope displacements to progressive failure were calculated, then they were analyzed and applied to criteria of displacements sensing.

To link the slope stability analysis with the maintenance method based on the measured data, the displacement until the slope failure to the finite element model was analyzed and applied to the maintenance method. A flowchart of the integrated analysis method was presented and a case study was conducted. The slope stability analysis was performed by generating a finite element model and, the time-dependent deterioration of slope was quantified by applying the SRF to the strength parameter of the soil. The FOS was calculated by the stress analysis method (SAM) of the finite element model, and the behavior up until slope failure was analyzed by nonlinear static analysis with k_0 condition. Each displacement of slope depth until failure is calculated from slope surface by finite element method (FEM). Accumulated displacement curve is derived each displacement from according to the strength reduction factor (SRF). It can draw the velocity curve and inverse-velocity curve. The three curves derived from FEM are applied to the slope maintenance method. The each displacement of slope depth is applied to the SPC method, and, it is the criterion of inclinometer for slope insides. In addition, it estimates the abnormal behavior of slope. Also, accumulated displacement curve of slope surface until the failure is used to make the mathematical failure model. Next, a mathematical failure model of the slope was predicted using a cumulative displacement curve. And the time of the slope failure was predicted using an inverse-velocity curve and, compared the formulation of Fukuzono [1] and the existing case of slope failure.

2. Theoretical background for stability analysis and maintenance of cut-slope

2.1 Finite element method (FEM) for slope stability analysis

The limit equilibrium method (LEM), a deterministic method, compares the shear stress and shear strength applied to assumed the failure surface of slope to present the FOS. However, because the limit equilibrium analysis provides only the minimum of FOS as the analytic result, it cannot present appropriate safety factor applicable to the analysis of progressive slope failure attributable to concurrent continuous displacement induced by a time-dependent deterioration of soil.

The finite element method can provide the measurement system for the maintenance of slope with proper analytic results. It also is an appropriate method to analyze the behavior of progressive slope failure [29]. The progressive slope failure has been examined through the finite element method. Zeinkiewicz, Humpheson and Lewis [8] had presented the strength reduction method that employed the SRF by which the safety factor was calculated as in the case of limit equilibrium analysis. Thereafter, Griffiths [10] had applied the strength reduction method to his analysis of progressive behavior of slope according to diverse soil conditions and geometries and verified the analytical method through comparative analyses with the chart (s) presented by Bishop and Morgenstern [30].

The SAM is a combination of the advantages of simple limit equilibrium method (LEM) and of finite element method (FEM), the advanced method. The stress state in slope is analyzed through finite element analysis, and the FOS of virtual active surfaces of the limit equilibrium analysis are calculated. Thereafter, the minimum of FOS and critical section among active surfaces in the limit equilibrium analysis are calculated. In the finite element analysis, the model of the material constituting the soil will use the Mohr-coulomb yield criteria identical to the failure criteria used in the limit equilibrium analysis.

The FOS of the slope to be determined by the stress analysis method is as expressed in the following Eq. (1).

$$\text{FOS} = \frac{\int_S \tau_f dT}{\int_S \tau_m dT} \tag{1}$$

Here, FOS, τ_m and τ_f denote the factor of the safety, induced shear stress, and shear strength according to Mohr-Coulomb failure criteria, respectively.

The FEM can analyze the slope behavior until failure, and if the SRF is applied, it also can quantify the time-dependent deterioration of slope. However, until now the FEM is not used for analyzing and comparing to the measured displacement data of slope.

2.2 Statistical process control method for detecting abnormal behavior of slope

The representative method used for the domestic maintenance of slope is the statistical process control (SPC) method broadly employed in manufacturing industries to reduce the level of defects of commercial products. The population for the SPC for the maintenance of slope is a set of measurements of the displacement of slope. The values of the mean and standard deviation of the population is used to judge the stability of slope statistically according to respective values plotted on the region beyond or within the control limit determined by the control chart. By using the control chart, the anomaly in the stability of slope can be found easily.

The method above uses the $\overline{X}$ control chart (of mean values) and the R control chart (of standard deviation, the varying range of measurements) simultaneously.

The control limit and centerline of the$\overline{X}$ control chart can be obtained by using the Eq. (2) represented in the following.

$$\begin{aligned} \text{UCL} &= \bar{\bar{x}} + A_2\overline{R} \\ \text{CL} &= \bar{\bar{x}} \\ \text{LCL} &= \bar{\bar{x}} - A_2\overline{R} \end{aligned} \tag{2}$$

Where, the constant A_2 is a function of sample sizen. $\bar{\bar{x}}$ is an estimation of μ, the mean of the population; and $\overline{R}$ represents an estimation of σ, the standard deviation of the population.

The control limit and centerline of the R control chart can be obtained by using the Eq. (3) represented in the following.

$$\begin{aligned} \text{UCL} &= D_4\overline{R} \\ \text{CL} &= \overline{R} \\ \text{LCL} &= D_3\overline{R} \end{aligned} \tag{3}$$

Where, the constants D_3 and D_4 are the functions of the sample size n.

KICT [20] applied the measured data of real slope to SPC method. After that, Yoo [22] proposed a statistical decision algorithm for maintenance of the slope based on the measured data. Therefore, it is necessary to calculate the management criteria of statistical process control method by integrating it with the finite element analysis which can simulate the failure of the slope.

2.3 Prediction method of the time of slope failure

To predict the time of the slope failure, diverse models employed the correlation between the time and the displacement varying according to the creep behavior of bedrock have been proposed [25, 31]. Since such models were difficult to apply to the prediction of the slope failure as a generalized model, Fukuzono [1] proposed the model of inverse-velocity expressing the changing characteristics of ground displacement as the relationship between time and inverse-velocity. The proposed inverse-velocity model was derived by using the measurements of varying acceleration obtained from the large-scaled actual experiment simulated an artificial landslide. Fukuzono [1] proposed the time of the slope failure as expressed in the following Eq. (4) by taking the trend line of inverse-velocity approaching 0 in accordance with the increasing velocity of displacement into account.

$$\frac{1}{V} = [A(\alpha - 1)]^{1/\alpha - 1} \bullet (t_f - t)^{1/\alpha - 1} \tag{4}$$

Here, t, t_f, and V denote the time, time of failure, and displacement velocity; and A and α are the constants introduced for curve fitting. When the value of α is bigger than 2, the shape of curve is convex otherwise the shape of curve would be concave with the value of α less than 2.

To exploit the advantage of the value of inverse-velocity becoming 0 at the time of the slope failure, the relevance between time and inverse-velocity has been widely used [32, 33]. Petley, Bulmer and Murphy [32] have analyzed the patterns of trend lines of the cases the slope failure however, since the patterns were analyzed from the measurements of the displacement after the failure of faces of slopes, it would be very difficult to get actual real-time prediction of the time of failure. Therefore, an analysis on the pattern of inverse-velocity curve at the design stage is needed to predict the time of the slope failure in the stage of performing maintenance works based on measurements. Thus, in this study, the procedure to analyze the progressive slope failure was presented in a way to link the procedure to the maintenance methods based on actual site measurements.

3. Integrated analysis methods of slope stability analysis and maintenance

3.1 Introduction

Figure 1 is an integrated analysis method for stability analysis and maintenance of cut-slopes.

The integrated analysis method presented in this study is divided into 14 steps. It is divided into three sections. In the first section, the geometric of the slope and the finite element model of FEM are generated. And, to consider the time-dependent deterioration, the SRF is applied to the strength parameter of the soil (step 1 ∼ 3). The integrated analysis method proposed in this study is based on the cut-slope and only the strength degradation caused by the time-dependent deterioration is taken into consideration. The proposed method can be used both as soil and rock as a material of cut slope, but the only soil is considered in this study.

In the second section, slope stability analysis is performed using FEM. The FOS is calculated by the SAM and the behavior up to the slope failure is analyzed by the nonlinear static analysis withk_0 condition. The displacements until the slope failure analyzed using the slope stability analysis plot the cumulative displacement curve,

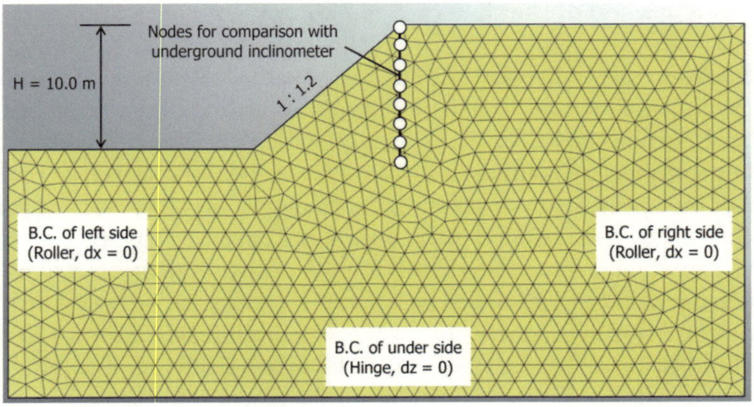

Figure 1.
Mesh and boundary conditions of slope stability analysis.

velocity curve, and inverse-velocity curve and, applied to the maintenance methods of the slope. The strength degradation of the soil due to the time-dependent deterioration phenomenon was quantified by SRF. However, geological factors and strength degradation due to groundwater cannot be considered. Therefore, the proposed method in this study cannot be applied when the groundwater level is located on the predicted failure surface of the slope.

In the last section, the results of the slope stability analysis are applied to the maintenance method. A typical sensor used for slope maintenance is an in-site inclinometer, which is applied to SPC using slope stability analysis results at the same point. Next, a mathematical failure model of the slope was predicted using a cumulative displacement curve. And the time of the slope failure was predicted using an inverse-velocity curve and, compared the formulation of Fukuzono [1]. Finally, the collapse behavior of Selborne in the United Kingdom and Kunini Slope in Japan, as reported by Petley [34], was compared.

Figure 2 in this study improves the problem of the slope stability analysis method that only estimates the FOS and, the problem of maintenance method that fails to evaluate clear management criteria and failure behavior of slope. It also combines the advantages of slope stability analysis and maintenance method based on measured data [35, 36]. Slope stability analysis is used to determine the failure behavior and FOS of the slope. And, it can be applied to the maintenance method to predict the management criteria of statistical process control method, a mathematical failure model of slope, and the time of slope failure.

In the maintenance of the slope, the surface displacement is usually tension wire, and the ground displacement is the inclinometer. Because the displacement of the entire slope is analyzed using the FEM, that can be applied to the management criteria of the displacement measuring instrument installed on the slope. In particular, the displacement at the crown of the slope is very important, which can be measured both by tension wire and inclinometer. There are two maintenance methods presented in this study. First, the failure model is calculated by the cumulative displacement curve of the slope. And the displacement according to the depth is applied to the SPC method to judge the occurrence of abnormal behavior of the slope. If only tension wire is applied to the maintenance of the slope, the formation of the failure surface cannot be confirmed because only the surface displacement of the slope can be measured. Because the inclinometer measures the displacement of the whole underground, it has an advantage that it can easily judge the failure surface. Therefore, the inclinometer was applied to the slope instrument in this study. Please refer to KICT [20] for the advantages and disadvantages of both measurement devices.

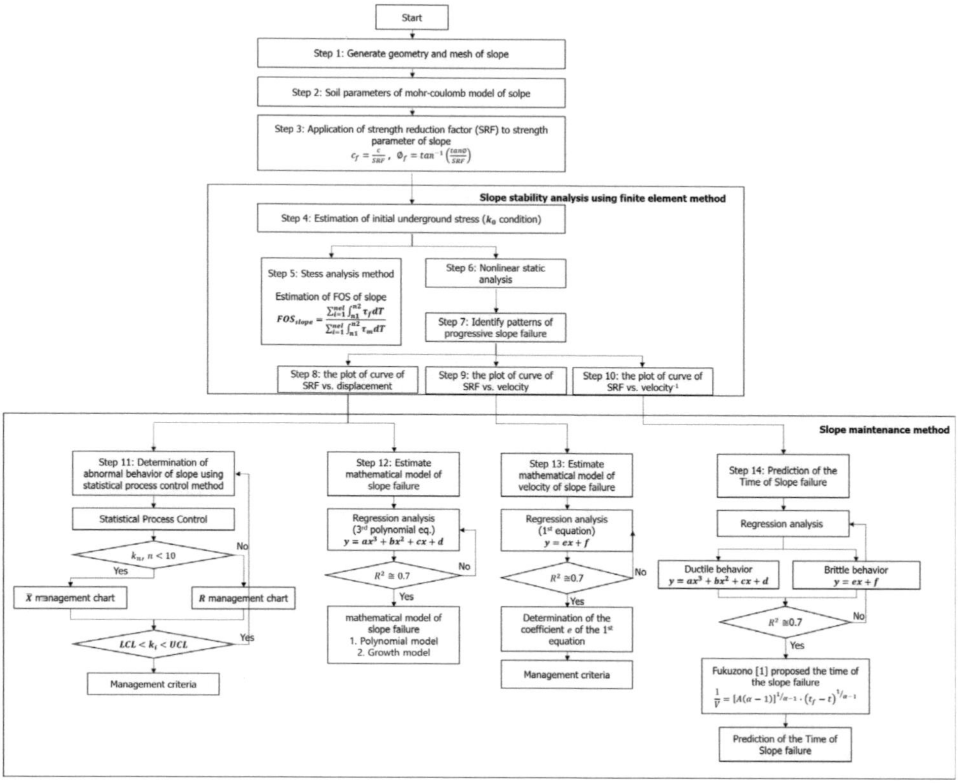

Figure 2.
A flowchart on integrated analysis methods of slope stability analysis and maintenance of the slope.

3.2 Analysis procedure

In this section, detailed description is given for each step according to the flowchart shown in **Figure 2**. The details are as follows:

1. Step 1: This is the step of generating the geometry and mesh of the slope. And the load and boundary conditions acting on the slope. In general, only self-weight is considered. Please refer to the literature on FEM [10, 11, 29].

2. Step 2: In the FEM, the soil uses the Mohr-coulomb model. The soil investigation is performed to calculate the cohesive and internal friction angle, which are the strength parameters of the soil.

3. Step 3: The strength parameters of soil are to be decreased according to the SRF, and the cohesion of soil at the point of failure and the internal friction angle are as expressed in Eqs. (5) and (6), respectively. In the case where the SRF is used, the value of initial strength reduction factor (SRF) and the value of increment should be determined. The values of 1.0 and 0.05 of the strength factor and increment are frequently applied to the method of general strength reduction. The final strength reduction coefficient will be a value at the stage failed to converge in the nonlinear static analysis.

$$c'_f = \frac{c'}{SRF} \tag{5}$$

$$\phi'_f = \tan\left(\frac{\tan\phi'}{SRF}\right)^{-1} \tag{6}$$

where, c′, $c_f{}'$, ϕ′, and $\phi_f{}'$ denote the cohesion, cohesion at the point of failure, internal friction angle, and internal friction angle at the point of failure, respectively. SRF denote the strength reduction factor.

4. Step 4: To conduct the finite element analysis, the calculation of initial underground stress of slope is important and, the value is calculated under the k_0 condition based on the data obtained from subsurface investigation.

5. Step 5: Based on the calculated initial underground stress of the slope, the FOS is calculated through the SAM [37]. In this case, the decrease in the value of safety factor is identified through an iterative analysis to be conducted according to the reduced strength parameter, and the failure surface will be applied to it as in the case of limit equilibrium analysis.

6. Step 6: The nonlinear static analysis will be conducted to analyze the behaviors of progressive slope failure. The iterative analysis should be continued until the value fails to converge (=until the occurrence of failure) according to the stages of each strength reduction factor.

7. Step 7: The displacement resulted from the progressive slope failure is identified through the iterative analysis conducted according to the strength reduction.

8. Step 8: After completing the calculation of cumulative displacement curve of the points of maximum displacement on slope through following the process in Step 7 above, the obtained results should be linked to the plan of the measurement of slope.

9. Step 9: The failure model of slope is created by using the velocity curve made from the analytic results and then it will be compared with the cumulative displacement curve made from the measurements collected through the stages of maintenance of the slope.

10. Step 10: In this step, plot an inverse-velocity curve for predict the time of slope failure.

11. Step 11: From this stage, the results of the slope stability analysis are applied to the maintenance method. In step 11, the displacement calculated at the point where the in-place inclinometer is installed is applied to the statistical process control method. And judges whether the management criteria is exceeded by depth of slope. Finally, determine whether the slope failure behavior is occurring and the location of the failure surface.

12. Step 12: In this step, we calculate the mathematical failure model of the slope. The calculated mathematical failure model is used as a reference for the displacement results measured by the in-place inclination.

13. Step 13: In this step, first order linear equation is calculated by regression analysis of velocity curve of the slope. It is difficult to derive a clear engineering meaning like the accumulated displacement curve of step 12 and

the inverse-velocity curve of step 14. Therefore, if the slope failure happens, the measured displacement velocity curve changes into the 1st-degree polynomials. It means the slope displacement velocity turns into invariable velocity, and it is estimated that the slope failure happens. Therefore it could be important criterion for slope maintenance.

14. Step 14: This stage predicts the time of slope failure in slope maintenance. Curve fitting is performed by regression analysis of the inverse-velocity curve by slope stability analysis. According to the existing literature, slopes with ductile behaviors are a third polynomial equation and slopes with brittle behavior are a linear equation. Regression analysis results are also compared with Fukuzono [1]'s slope failure prediction formula.

4. Application result of the integrated analysis method

4.1 Generate finite element model for slope stability analysis (step 1 $\sim$ 10)

In the first section, the geometric of the slope and the finite element model of FEM are generated. And, to consider the time-dependent deterioration, the SRF is applied to the strength parameter of the soil (step 1 $\sim$ 3).

4.1.1 Step 1: Generate geometry and mesh of slope

For the slope stability analysis of the progressive behavior of slope, the geometry and meshes of the finite slope of 10.0 m in height and 1: 1.2 of standard slope were created (**Figure 1**). For this research, the cut-slope modeling is obliged the standard height and incline suggested by the design manuals [19, 38]. The displacements until progressive failure analyzed FEM are compared to the measured data, the criteria come from the inclinometer data.

The boundary condition of left and right side is dx = 0 and same as roller. The boundary condition of the floor is dz. = 0, which is the same as the hinge. The load applied only its own gravity. In the finite element model, the element at the point where the in-place inclinometer is installed should be identified.

4.1.2 Step 2: Soil parameters of Mohr-coulomb model of solpe

The Mohr-coulomb model was selected for the FEM for which the internal friction angle($\varnothing'$), cohesion (c'), dilatancy angle (ψ), Poisson's ratio (υ), and unit weight (γ_t) are needed. Among such factors, the relative importance of the dilatancy angle that reflects the change in volume resulting from the yield process is less significant in the analysis of the stability of slope that calculates the safety factor [11]. In this study, the value of dilatancy angle was set 0° to let the change in volume to be constantly applied to the analysis. The values of the physical properties of the soil are as summarized in **Table 1**.

The soil slope behaviors are largely divided into two categories; firstly, ductible behavior in case of small particle soils, at second, brittle behavior in case of coarse particle soils. Mohr-coulomb model is used to analyze those two cases, which is the elastic-perfect plastic model. The case 1 is for the ductile behavior of slope, whose cohesion value is applied as 10 kPa to show the behavior small soil particles. The case 2 is for the brittle behavior of slope, whose cohesion value is 0 kPa to show the brittle slope behavior of coarse soil particles.

The soil of Case 1 has a cohesive of 10 kPa and an internal friction angle of 30 degrees. The soil of Case 2 has a cohesive of 0 kPa and an internal friction angle of 40 degrees. The slope of Case 1 shows the ductility behavior due to the cohesive of soil, and the slope of Case 2 shows the brittle behavior because the soil has no cohesive and the internal friction angle is large. The analytical results are compared according to the material characteristics of these slopes.

4.1.3 Step 3: Application of strength reduction factor to strength parameter of slope

In step 3, the time-dependent deterioration of slope was quantified by applying the SRF to the strength parameter of the soil. The strength parameters of the slope shown in **Table 1** were reduced according to the SRF as shown in **Figure 3**. The slope stability analysis is performed with the reduced strength parameters, and iterative analysis is performed until it is not converged. If it does not converge, the slope has collapsed in the finite element analysis. Until now, the modeling is identical to that of the usual strength reduction factor (SRF) method of slope stability analysis. [8, 39, 40].

4.2 Slope stability analysis (step 4 $\sim$ 10)

In the second section, and slope stability analysis using finite element model. The FOS is calculated by the stress analysis method (SAM) and the behavior up to the slope failure is analyzed by the nonlinear static analysis withk_0 condition. The

Parameter	Case 1	Case 2
$\gamma_t (kN/m^3)$	19.0	19.0
E (kPa)	40,000	40,000
υ	0.28	0.28
$\varnothing'$ (°)	30.0	40.0
c' (kPa)	10.0	0.0
ψ (°)	0.0	0.0

Table 1.
Physical parameters of the soil of slope applied to slope stability analysis.

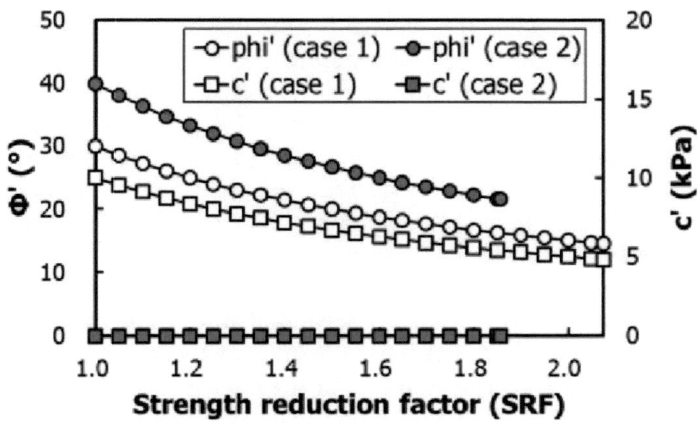

Figure 3.
Variation of the strength parameters of slope according to strength reduction factor.

displacements until the slope failure analyzed using the slope stability analysis plot the cumulative displacement curve, velocity curve, and inverse velocity curve and, applied to the maintenance methods of the slope.

4.2.1 Step 4 ~ 7: Conduct the slope stability analysis of finite element model and estimation of behavior until the slope failure

In this step, slope stability analysis is performed using the finite element model generated in the first section. In slope stability analysis using FEM, it is very important to estimate the stress distribution in the slope. In Step 5, the initial stress distribution of slope is estimated at k_0 according to the coefficient of earth pressure [37]. Then, the FOS of the slope is calculated by the SAM. As shown in **Figure 3**, the strength parameters according to the SRF is applied, and the FOS of the slope by the SAM is shown in **Figure 4(a)** for case 1 and **Figure 5(a)** for case 2. In Step 6, nonlinear static analysis is performed by gradually reducing the strength parameters of the slope. The nonlinear static analysis is repeated until the analysis is not converged. If the nonlinear static analysis is not converged, the slope has collapsed. As a result of the finite element analysis, the soil failure occurred at SRF 2.075 in Case 1 and SRF 1.856 in Case 2.

Figure 6 is the distribution of displacement and shear strain by finite element analysis under slope failure condition in the slope of case 1. The legend in **Figure 6(a)** shows the amount of displacement, and in **Figure 6(b)** shows the effective shear strain. When the slope stability analysis is performed by the proposed method, the behavior up to the progressive failure of the slope can be analyzed. The distribution of shear strain can predict the failure surface of the slope. One of the greatest advantages is that the failure surface can be estimated by the finite element analysis in the slope consist of the continuous soil. Step 7 is the same as the general procedure for slope stability analysis using finite element analysis. In this study, only 2D finite element analysis was performed. Three-dimensional analysis is also possible. For finite element analysis using the SRF, see the paper by Wei, et al. [40].

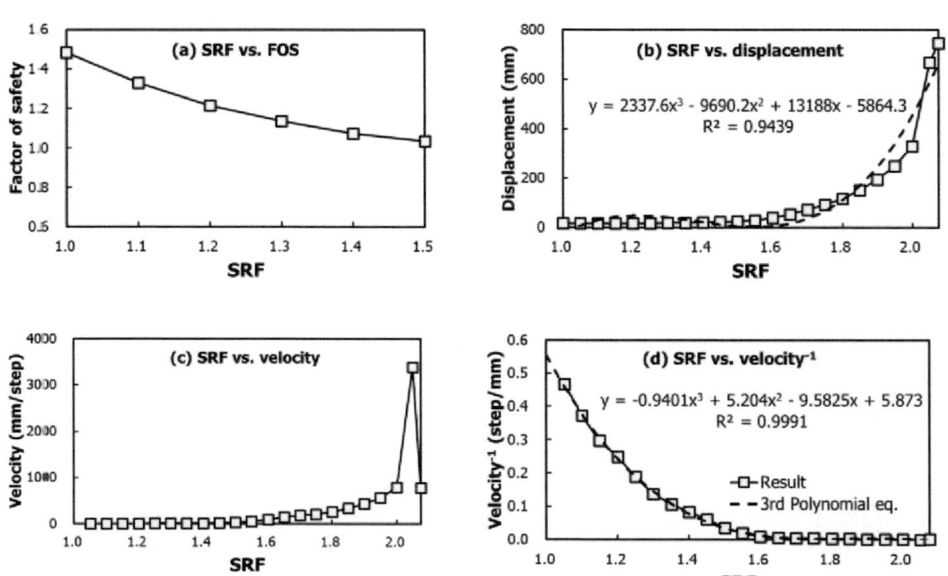

Figure 4.
The results of slope stability analysis in the slope of case 1; (a) factor of safety, (b) SRC vs. displacement curve, (c) SRC vs. velocity curve, (d) SRC vs. inverse velocity curve.

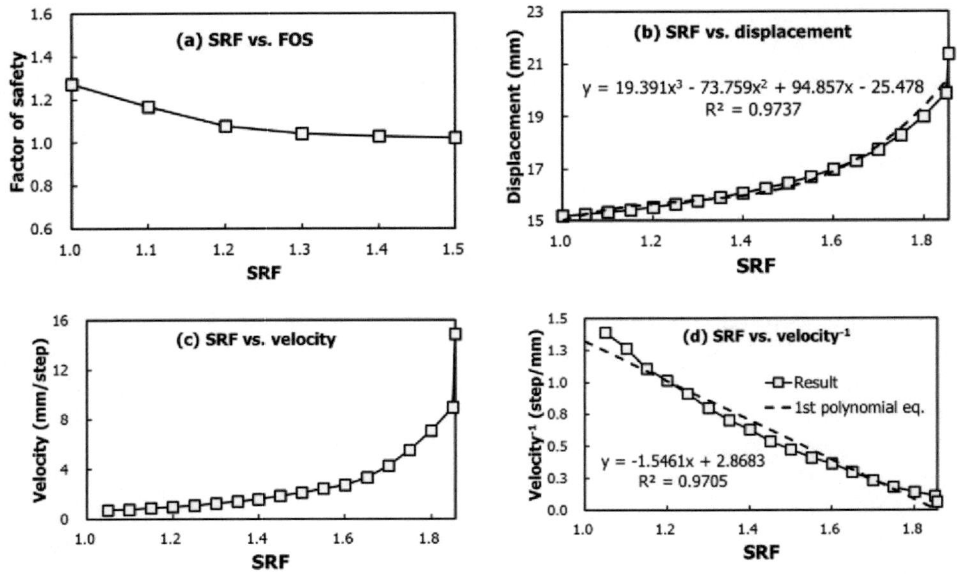

Figure 5.
The results of slope stability analysis in the slope of case 2; (a) factor of safety, (b) SRC vs. displacement curve, (c) SRC vs. velocity curve, (d) SRC vs. inverse velocity curve.

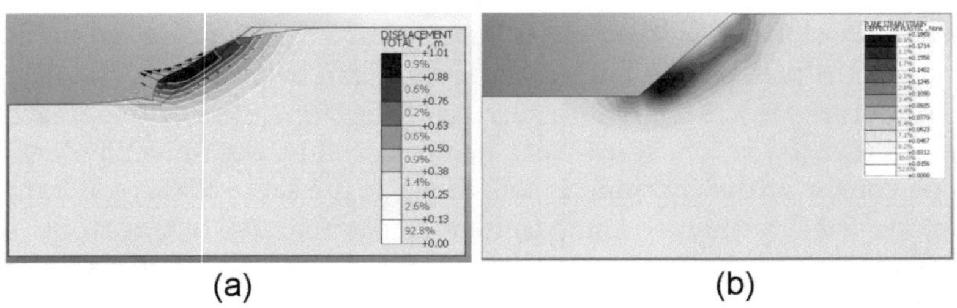

Figure 6.
Distribution of displacement and shear strain by finite element analysis under slope failure condition (case 1).

4.2.2 Step 8 ~ 10: Nonlinear static analysis of slope finite element model and summarization of results

In the third stage, the displacements until the slope failure analyzed using the slope stability analysis plot the cumulative displacement curve, velocity curve, and inverse velocity curve and, applied to the maintenance methods of the slope. In the finite element analysis, the displacements of all nodes are calculated in the finite element model of the slope. Here, only the displacement of the node at the point where the displacement meter such as an inclinometer or a tension wire is installed is used. In particular, the displacement at the crown of the slope is most important. Therefore, in this stage, a graph was created by displacement at the crown of the slope.

Figure 4 shows the result of slope stability analysis in the slope of case 1. Above all, the FOS was calculated by using the stress analysis method and then, the cumulative displacement curve, the displacement velocity curve, and the displacement inverse-velocity curve was plotted to illustrate the displacement of the upper part of the slope. **Figure 4(a)** Represents the reduction in safety factor resulted from the stress analysis. The FOS was 1.0 when the SRF was 1.0 and the FOS reached 1.0 when the SRF was 1.5. The calculated displacement at each stage of reduced strength is as illustrated in **Figure 4(b)–(d)**. In the cumulative displacement curve plotted in **Figure 4(b)**, the rapid progression of displacement, initiated at the stage of 1.6 of strength reduction factor and then evolved rapidly after the stage of 2.0 of SRF, is illustrated. The failure model of the slope was calculated as in the following 3rd order polynomial equation:$y = 2337.6x^3 - 9690.2x^2 + 13188x + 5864.3$. In **Figure 4(d)**, the inverse-velocity curve is estimated as a third order polynomial of $y = -0.9401x^3 + 5.204x^2 - 9.5825x + 5.873$.

Figure 5 shows the result of slope stability analysis in the slope of case 2. **Figure 5(a)** shows the change of FOS according to SRF. Compared with case 1, the decreasing slope was more moderate. When SRF was 1.0, FOS was 1.27, which was smaller than case 1. When SRF was 1.5, FOS reached 1.0 and was the same as case 1. In the cumulative displacement curve plotted in **Figure 5(b)**, compared to that in the case 1 (**Figure 4(b)**), the displacement proceeded continuously from the initial stage and the displacement increased rapidly after the stage of 1.6 of the strength reduction factor. The failure model of the slope was calculated as in the following 3rd order polynomial equation:$y = 19.391x^3 - 73.759x^2 + 94.857x - 25.478$. In **Figure 5(d)**, the inverse displacement velocity curve was calculated as a 1st linear equation of$y = -1.5461x + 2.8683$, unlike case 1. The inverse displacement velocity curve is used to predict slope failure time [1, 24]. The inverse displacement velocity curves of case 1 and case 2 were compared at step 14.

4.3 Slope maintenance method (step 11 $\sim$ 14)

In the last section, the results of the slope stability analysis are applied to the maintenance method. For the maintenance of the slope, the displacement of the slope is measured by tension wire or inclinometer. However, no technique has been proposed to determine management criteria. The typical sensor used for slope maintenance is an in-site inclinometer, which is applied to statistical process control using slope stability analysis results at the same point. The displacement at each depth measured in the inclinometer can be calculated by applying the SPC method to the upper and lower control limits (UCL, LCL). If the displacement exceeds this management criterion, an abnormal behavior has occurred. Next, a mathematical failure model of the slope was predicted using a cumulative displacement curve. The cumulative displacement data of the slope measured over time are compared with the behavior up to the failure estimated by the finite element analysis. This can qualitatively determine whether the slope is causing the failure behavior. And the time of the slope failure was predicted using an inverse velocity curve and, compared the formulation of Fukuzono [1].

The inverse-velocity curve is a method of predicting the time of failure of the slope. If the measured inverse-velocity curve shows this pattern, it can be predicted the time of slope failure. Finally, the collapse behavior of Selborne in the United Kingdom and Kunini Slope in Japan, as reported by Petley [34], was compared.

4.3.1 Step 11: Determination of abnormal behavior of slope using statistical process control method

In step 11, the displacement calculated at the point where the in-place inclinometer is installed is applied to the statistical process control method. In the finite element analysis, the displacement of the entire slope is estimated. Depending on the SRF, the displacement at the point where the inclinometer is installed can be estimated. The SPC method determines the abnormal behavior at the point where the inclinometer is installed, as described in Section 2.2. This abnormal region appears when a failure surface is formed due to the progressive behavior of the slope. SPC method can be statistically evaluated and have the advantage of setting upper and lower control limits. Finally, determine whether the slope failure behavior is occurring and the location of the failure surface.

The in-place inclinometer measures the displacement of the ground at intervals of 1.0 m. From the results of the slope stability analysis, the displacements at each depth were analyzed at intervals of 1.0 m as shown in **Figure 1**. The calculated displacement is applied to the statistical process control method as shown in **Figure 7**. The depth of the horizontal axis shows the depth from the ground surface. The vertical axis shows the displacement of the slope. **Figure 7(a)** shows that when the strength reduction factor is 1.6, **Figure 7(b)** is 1.7, **Figure 7(c)** is 1.8, **Figure 7(d)** is 1.9, **Figure 7(e)** is 2.0 and **Figure 7(f)** is 2.075. **Figure 7(a)** shows the case where the strength reduction factor is 1.6, and the displacement at all depths does not exceed the management criteria of upper control limit. **Figure 7(d)** shows that the strength reduction factor is 1.9 and exceeds the management criteria of upper control limit. From this time, a failure surface of the slope was formed. By applying the SPC method, it is possible to judge the failure behavior of the slope and the generation of the failure surface.

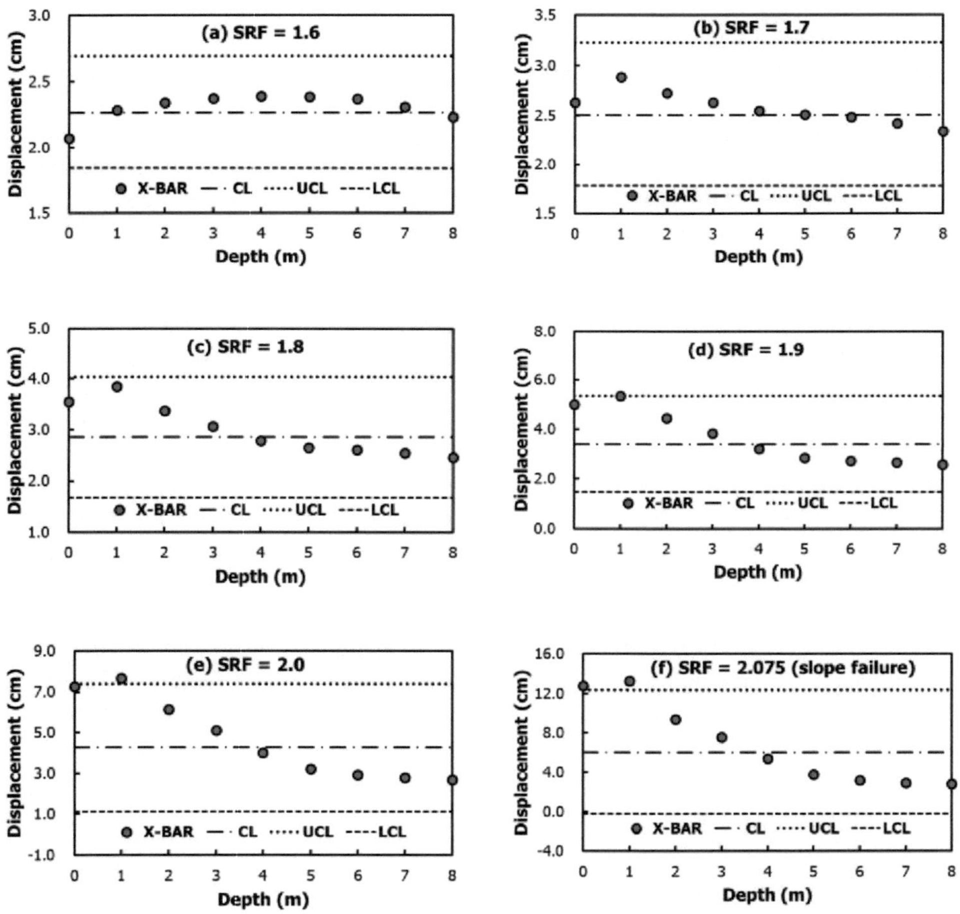

Figure 7.
$\overline{X}$ management chart according to depths (Case 1); (a) SRF = 1.6, (b) SRF = 1.7, (c) SRF = 1.8, (d) SRF = 1.9, (e) SRF = 2.0, SRF = 2.075 (slope failure).

4.3.2 Step 12: Estimate mathematical model of slope failure

In this Step 12, a mathematical model of slope failure is predicted by the displacement result of the slope stability analysis. As shown in Step 11, the displacement up to the slope failure at the point where the in-place inclinometer is installed is used.

Figure 8 shows the mathematical failure model using the cumulative displacement curves for each depth. The largest displacement occurs on the ground surface, and the displacement is hardly generated below the failure surface of the slope. The location of the failure surface can be determined in step 4 through the slope stability analysis, as shown in **Figure 6(b)**. As shown in **Figure 8(f)**, little displacement occurred at the 5.0 m depth, because it exists below the failure surface of the slope.

The cumulative displacement curve through slope stability analysis can be applied as a management criterion for the measured data of the in-place inclinometer.

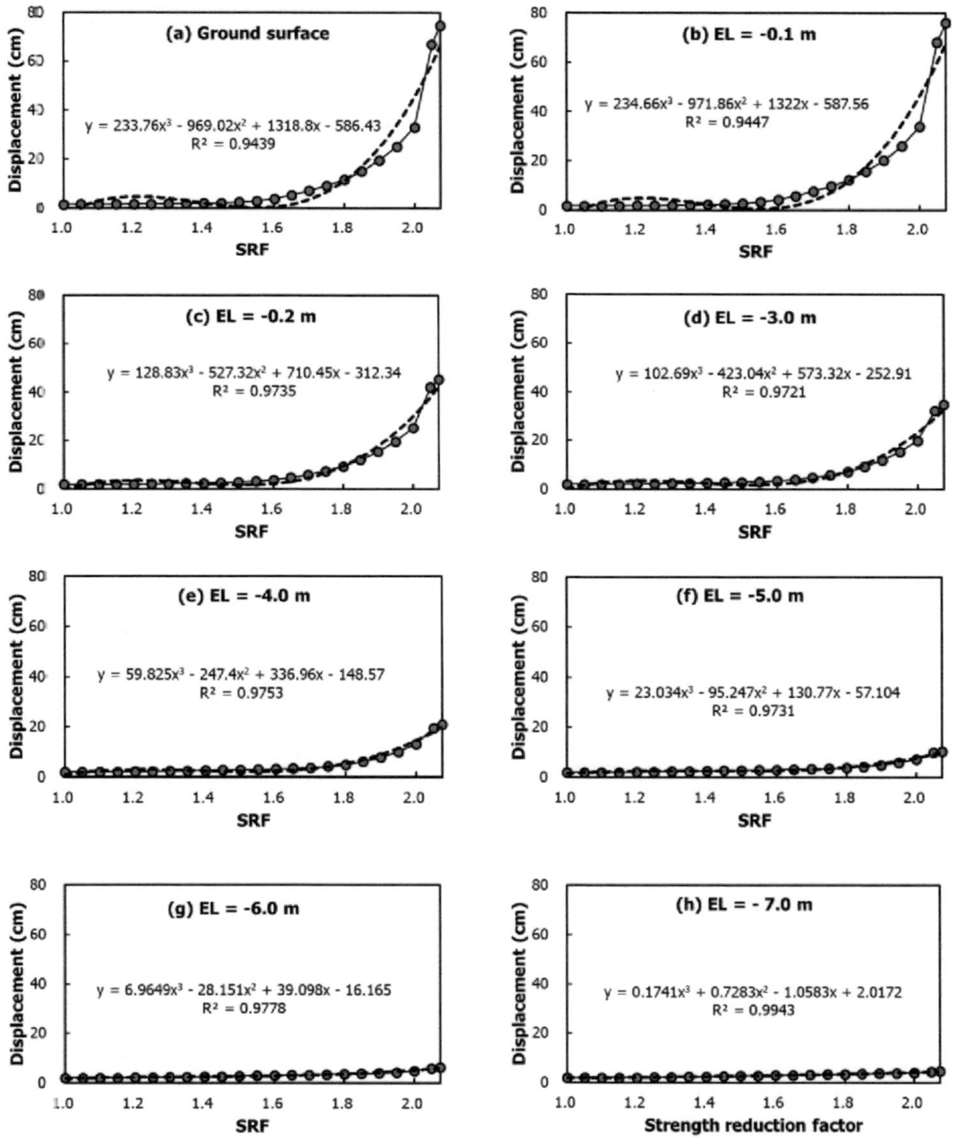

Figure 8.
The mathematical failure model using the cumulative displacement curves for each depth (case 1).

4.3.3 Step 14: Prediction of the time of slope failure

This stage predicts the time of slope failure in slope maintenance. Curve fitting is performed by regression analysis of the inverse-velocity curve by slope stability analysis. According to the existing literature, slopes with ductile behaviors are a third polynomial equation and slopes with brittle behavior are a linear equation. Regression analysis results are also compared with Fukuzono [1]'s slope failure prediction formula.

Fukuzono [1] intended to predict the time of slope failure resulted from progressive behavior with the inverse-velocity curve. The resulted inverse-velocity curves are as illustrated in **Figures 9** and **10**. In the case of the soil strength parameters of the cohesion (10kPa) and internal friction angle (30°) in the Case 1, the displacement inverse-velocity was reduced rapidly to the 3rd order polynomial equation and, the values of $A = 4.0$ and $\alpha = 1.21$ resulted from the equation presented by Fukuzono [1] that showed a convex pattern of the curve (**Figure 9 (a)**). The pattern was similar to the pattern of ductile behavior in the case of 'The Collapse of Kunini Slope' in Japan (**Figure 9(b)**).

And in the case of the soil strength parameters of the cohesion (0kPa) and internal friction angle (40°) in the Case 2, the displacement inverse-velocity was

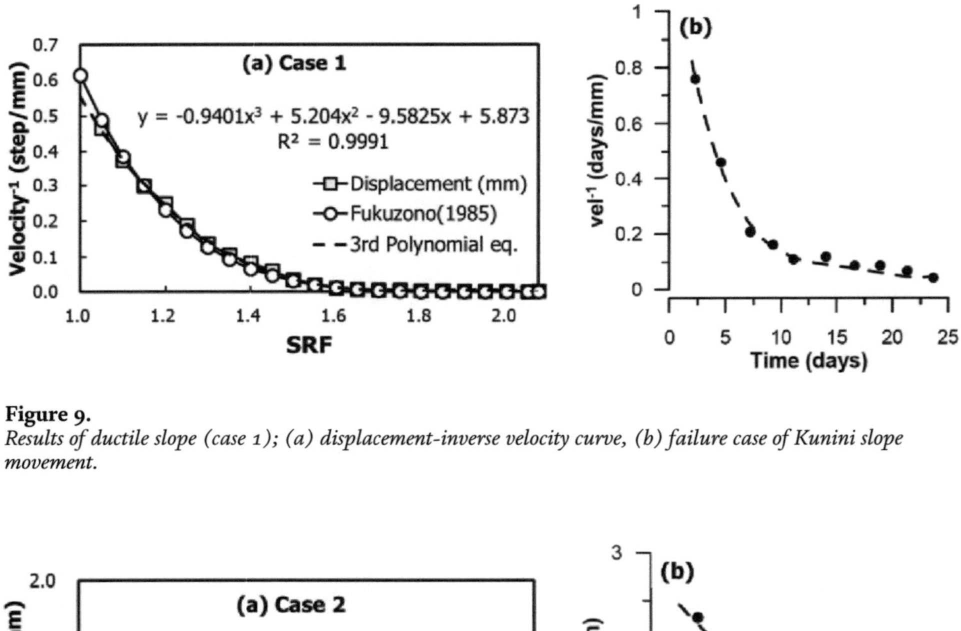

Figure 9.
Results of ductile slope (case 1); (a) displacement-inverse velocity curve, (b) failure case of Kunini slope movement.

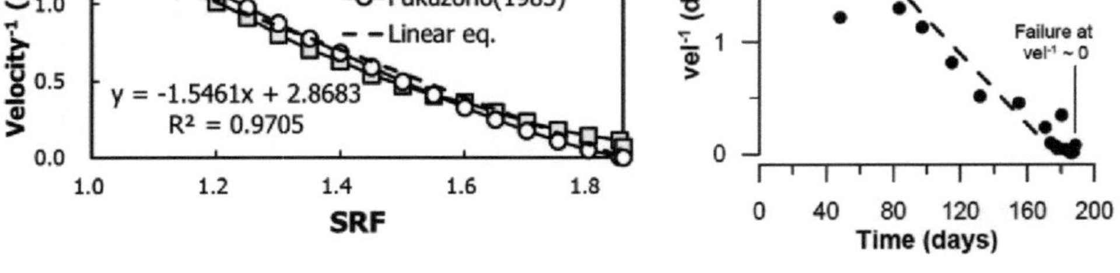

Figure 10.
Results of brittle slope (case 2); (a) inverse displacement velocity curve, (b) failure case of Selborne cut-slope experiment.

reduced to the 1st order linear equation and rendered values of A = 2.05 and α = 1.79 showing linear pattern of the curve (**Figure 10(a)**). The pattern was similar to that in the case of 'The Failure of the Cut Slope of Selborne' in the United Kingdom reported by Petley [34] (**Figure 10(b)**).

5. Discussion

This study developed an integrated analysis method for stability analysis and maintenance of cut-slopes in urban. To link the slope stability analysis with the maintenance method based on the measured data, the displacement until the slope failure to the finite element model was analyzed and applied to the maintenance method. The integrated analysis method proposed in this study is based on the cut-slope and only the strength degradation caused by the time-dependent deterioration is taken into consideration. The proposed method can be used both as soil and rock as a material of cut slope, but the only soil is considered in this study.

The integrated analysis method in this study can complement the disadvantages of the slope stability analysis and integrate it with the maintenance method based on the measured data of slope. The slope stability analysis can be used to quantify the displacement until slope failure as the cumulative displacement curve, velocity curve, and inverse velocity curve. The results of slope stability analysis could be used as management criteria for statistical process control method, mathematical model and the time of slope failure applied to maintenance. Then, the failure behavior of the slope and the generation of the failure surface were confirmed. The displacement of the slope analyzed by the finite element analysis should be the same as the position of the displacement meter installed on the slope.

By the comparison of this model with the failure model based on measured data, the obtained failure model was concluded as a 3rd order polynomial failure model equivalent to that of the site of 'Neureupjae' presented by Han and Chang [41]. **Figure 4(c)** represents the velocity curve. It also shows the rapid progression of displacement velocity at the point of 2.0 of SRF. The corresponding displacement inverse-velocity curve is illustrated in **Figure 4(d)**. The inverse-velocity was generated by the following 3rd order equation,$y = -0.9x^3 + 5.2x^2 - 9.6x + 5.9$ that rendered the rapid decrease in the inverse-velocity.

The behavior of the slope appeared almost identical with that in the Case 1 and, the 3rd order polynomial equation similar to that of the site of 'Neureupjae' presented by Han and Chang [41] was also derived. The equation appeared in 3rd order polynomial equation: $y = 19.4x^3 - 73.8x^2 + 94.9x - 25.5$. **Figure 5(c)** represents the displacement velocity curve of cumulative displacement on which there are two points of inflection at each point of 1.6 and 1.9 of the strength reduction factor. **Figure 5(d)** shows the displacement inverse-velocity curve. Contrary to that in the Case 1, the equation was reduced to the 1st order linear one:$y = -1.6x + 2.9$.

The slope stability analysis conducted with conditions defined in the Case 1 rendered following results of the changes in initial soil strength (cohesion) and internal friction varied from 10kPa to 4.8kPa and from 30° to 14.5°. The mathematical model of slope failure by cumulative displacement curve was reduced to the 3rd order polynomial equation,$y = 2337.6x^3 - 9690.2x^2 + 13188x + 5864.3$. The inverse-velocity curve resulted from the analysis of the Case 1 appeared as the 3rd order polynomial equation,$y = -0.9x^3 + 5.2x^2 - 9.6x + 5.9$. The equation presented by Fukuzono yielded values A = 4.0 and α = 1.21 that determined the convex pattern of the curve similar to the pattern of ductile behavior appeared in the case of 'The Collapse of Kunini Slope' in Japan.

The results obtained from the analysis of the Case 2 showed that the failure resulted from the initial soil cohesion of 0 kPa with the internal friction varied from 40° to 21.5°. The resulted failure model corresponded to this cumulative displacement curve was reduced to the 3rd order polynomial equation, $y = 19.4x^3 - 73.8x^2 + 94.9x - 25.5$. Contrary to the Case 1, the displacement inverse-velocity curve of the Case 2 showed the pattern of linear equation, $y = -1.6x + 2.9$, and the values of $A = 2.05$ and $\alpha = 1.79$ were obtained therefrom. The pattern of this curve was similar to that in the case of 'The Failure of the Cut Slope of Selborne' in the United Kingdom.

6. Conclusions

When the road is constructed by the development of the urban, it is essential to cut-slope. However, the slope has gradually degraded the strength of the soil due to the time-dependent deterioration, and progressive slope failure is induced. This study developed an integrated analysis method for stability analysis and maintenance of cut-slopes in urban. The slope stability analysis was performed using the finite element model, and the progressive slope failure by time-dependent deterioration was quantified by using the strength parameters of soil applying the SRF. The displacements until the slope failure by slope stability analysis were quantified by cumulative displacement curve, velocity curve, and inverse velocity curve and, applied to the slope maintenance method. The inverse-velocity curve applied to the prediction of the time of slope failure was regressed to the 1st linear equation in the brittle material and the 3rd polynomial equation in the ductile material. This is consistent with the proposed formula of Fukuzono [1] and also shows similar behavior to the existing failure case.

Recently, sensors and communication technologies for measuring the behavior of slopes have been dramatically developed. However, there is no technique or management method to interpret it. This study aims to solve these problems.

The proposed method integrates the existing slope stability analysis and maintenance techniques. It is difficult to obtain the measurement data of the collapsed slope, but the proposed technique should be verified through the failure of the model slope. And the behavior of ductile and brittle behavior appeared depending on the characteristics of the materials constituting the slope. Research on this part is also necessary. In the future, integrated analysis method should be improved through additional research. And it should be applied to cut-slope of urban to prevent disasters.

In this study, the proposed method was carried out only on homogeneous slopes consist of soil. In the future, research on the slope with the heterogeneous and anisotropic is needed. Also, the proposed method should be verified by applying it to the 3D rock slope which causes collapse along the joint of rock.

Acknowledgements

This work was supported by Seoul Institute of Technology (SIT) [(2020-AA-007) Development and Application Plan on Smart GNSS Technology for Monitoring Displacement of Urban Infrastructure] The content of this chapter is the rewriting of Dr. Park Min-cheol's doctoral dissertation [36] and a paper [35] already published in Korea.

Author details

Mincheol Park[1]*, Heuisoo Han[2] and Yoonhwa Jin[2]

1 Seoul Institute of Technology, Seoul, Republic of Korea

2 Kumoh National Institute of Technology, Gumi, Republic of Korea

*Address all correspondence to: mcpark@sit.re.kr

References

[1] Fukuzono, T. In *A new method for predicting the failure time of a slope*, Proceedings of the 4th International Conference and Field Workshop in Landslides, Tokyo, 1985; 1985; pp 145–150.

[2] Wieczorek, G. F., Landslide triggering mechanisms. *Landslides: Investigation and mitigation* **1996,** 247, 76–90.

[3] Varnes, D. J., Slope movement types and processes. *Special report* **1978,** 176, 11–33.

[4] Conte, E.; Silvestri, F.; Troncone, A., Stability analysis of slopes in soils with strain-softening behaviour. *Computers and Geotechnics* **2010,** 37, (5), 710–722.

[5] Lo, K. Y.; Lee, C. F., Stress analysis and slope stability in strain-softening materials. *Geotechnique* **1973,** 23, (1).

[6] Chen, Z.; Morgenstern, N.; Chan, D., Progressive failure of the Carsington Dam: a numerical study. *Canadian Geotechnical Journal* **1992,** 29, (6), 971–988.

[7] Sterpi, D., An analysis of geotechnical problems involving strain softening effects. *International Journal for Numerical and Analytical Methods in Geomechanics* **1999,** 23, (13), 1427–1454.

[8] Zeinkiewicz, O.; Humpheson, C.; Lewis, R., Associated and non-associated visco-plasticity in soils mechanics. *Journal of Geotechnique* **1975,** 25, (5), 671–689.

[9] Chowdhury, R., Discussion on stability analysis of embankments and slopes. *J. Geotech. Engng, ASCE* **1981,** 107, 691–693.

[10] Griffiths, D. V. Finite element analyses of walls, footings and slopes. University of Manchester, United Kingdom, 1980.

[11] Griffiths, D.; Lane, P., Slope stability analysis by finite elements. *Geotechnique* **1999,** 49, (3), 387–403.

[12] Park, M. C., Behavior analysis by model slope experiment of artificial rainfall. *Natural Hazards and Earth System Sciences* **2016,** 16, (3), 789–800.

[13] Tohari, A.; Nishigaki, M.; Komatsu, M., Laboratory rainfall-induced slope failure with moisture content measurement. *Journal of Geotechnical and Geoenvironmental Engineering* **2007,** 133, (5), 575–587.

[14] Tami, D.; Rahardjo, H.; Leong, E.-C., Effects of hysteresis on steady-state infiltration in unsaturated slopes. *Journal of Geotechnical and Geoenvironmental Engineering* **2004,** 130, (9), 956–967.

[15] Conte, E.; Donato, A.; Troncone, A., A simplified method for predicting rainfall-induced mobility of active landslides. *Landslides* **2017,** 14, (1), 35–45.

[16] Chen, H.-E.; Tsai, T.-L.; Yang, J.-C., Threshold of Slope Instability Induced by Rainfall and Lateral Flow. *Water* **2017,** 9, (9), 722.

[17] Anderson, S. A.; Sitar, N., Analysis of rainfall-induced debris flows. *Journal of Geotechnical Engineering* **1995,** 121, (7), 544–552.

[18] Oh, S.; Lu, N., Slope stability analysis under unsaturated conditions: Case studies of rainfall-induced failure of cut slopes. *Engineering Geology* **2015,** 184, 96–103.

[19] MLTM *Construction of slope design criteria*; Korea, 2011; pp 119–133.

[20] KICT, Development of tunnel portal slope stabilization technique and real-time monitoring system considering deterioration characteristics. **2006,** 1–406.

[21] KEC *Slope maintenance monitoring system*; Korea, 2009; pp 1–10.

[22] Yoo, B. A Study of failure analysis methods based on real-time monitoring data for landslide warning system. Kumoh National Institute of Technology, Korea, 2006.

[23] Hayashi, S.; Bowon, P.; Komamura, F.; Yamamori, T., On the Forecast of Time to Failure of Slope (II). *Landslides* **1988,** 25, (3), 11-16_1.

[24] Fukuzono, T., Recent studies on time prediction of slope failure. *Landslide News* **1990,** 4, (9), 9–12.

[25] Voight, B., A method for prediction of volcanic eruptions. *Nature* **1988,** 332, (6160), 125–130.

[26] Martin, D. C. Time dependent deformation of rock slopes. University of London, 1983.

[27] Saito, M. In *Forecasting time of slope failure by tertiary creep*, Proc. 7th Int. Conf on Soil Mechanics and Foundation Engineering, Mexico City, 1969; Citeseer: 1969; pp 677–683.

[28] Rose, N.; Hungr, O. In *Forecasting Potential Rock Slope Failure In Open Pit Mines Using the Inverse-velocity Method-Case Examples*, 1st Canada-US Rock Mechanics Symposium, 2007; American Rock Mechanics Association: 2007.

[29] Duncan, J. M., State of the art: limit equilibrium and finite-element analysis of slopes. *Journal of Geotechnical engineering* **1996,** 122, (7), 577–596.

[30] Bishop, A.; Morgenstern, N., Stability coefficients for earth slopes. *Geotechnique* **1960,** 10, (4), 129–153.

[31] Saito, M.; Uezawa, H. In *Failure of soil due to creep*, Fifth International Conference of Soil Mechanics and Foundation Engineering, 1961; 1961; pp 315–318.

[32] Petley, D. N.; Bulmer, M. H.; Murphy, W., Patterns of movement in rotational and translational landslides. *Geology* **2002,** 30, (8), 719–722.

[33] Wartman, J.; Malasavage, N. E. In *Predicting Time-to-Failure in Slopes from Precursory Displacements: A Centrifuge Experiment*, Geo-Congress 2013: Stability and Performance of Slopes and Embankments III, 2013; ASCE: 2013; pp 741–749.

[34] Petley, D., The evolution of slope failures: mechanisms of rupture propagation. *Natural hazards and earth system sciences.* **2004,** 4, (1), 147–152.

[35] Park, M.; Yoo, B.; Baek, Y.; Hwang, Y., Integral Method of Stability Analysis and Maintenance of Slope. *Journal of the Korean Geoenvironmental Society* **2016,** 17, (3), 27–35.

[36] Park, M. Integral Analysis of Slope Behavior according to Progresive Failure and Rainfall. Kumoh National Institute of Technology, Gumi, Republic of Korea, 2016.

[37] MIDAS, I. T., *GTS NX on-line manual*. MIDAS Information Technology Co., Ltd.: Korea, 2013.

[38] KGS, Structure foundation design standards specification. In Korean Geotechnical Society (in Korean): 2009.

[39] Cheng, Y.; Lansivaara, T.; Wei, W., Two-dimensional slope stability analysis by limit equilibrium and strength reduction methods. *Computers and Geotechnics* **2007,** 34, (3), 137–150.

[40] Wei, W.; Cheng, Y.; Li, L., Three-dimensional slope failure analysis by the strength reduction and limit equilibrium methods. *Computers and geotechnics* **2009,** 36, (1–2), 70–80.

[41] Han, H. S.; Chang, K. T., Predicting the failure of slope by mathematical model. *Journal of the Korean Geotechnical Society* **2005,** 21, (2), 145–150.

Chapter 9

Ecological Engineering Measures for Ravine Slope Stabilization and Its Sustainable Productive Utilization

Gaurav Singh, Raj Kumar, Dinesh Jinger and Dinesh Dhakshanamoorthy

Abstract

The 120 countries have committed to set the UNCCD sustainable development goal on achieving the land degradation neutrality by 2030 including India. The target has to be accomplished in a synergistic and cost-effective manner in accordance with countries' specific national contexts and development priorities. Globally, the ravine landscapes are considered among the world's most degraded ecosystems. Therefore, restoring ravines is considered a high priority item in the natural resource management programs. The vegetation cover augmented with appropriate conservation measures is the most sought restoration strategy. The engineering measures are prerequisite for slope stabilization and sustainable productive utilization in ravine ecosystem. The several methods for slope stabilization are available but only few are applied in ravine land. Therefore, in this chapter, we have covered only those slope stabilization techniques which were successfully applied for the restoration of ravine land.

Keywords: ravine, soil erosion, runoff, land degradation, slope stabilization

1. Introduction

The ravine landscape is considered as one of the most degraded terrestrial ecosystems in the world. The continuous and aggravated soil erosion and subsequent anthropogenic activities leads to the formation of ravine landscapes. Ravine is intricate network of large gullies with depth ranging from 3 m or more and affected by extreme form of mass soil erosion. The high drainage density and multi-directional steep slopes are other morphological characteristics of ravines. In India, ravines occupy about 4.3 M ha along the rivers *Yamuna*, *Chambal*, *Mahi*, and *Sabarmati* [1]. The marginal lands near to gully head in ravines are dominated by agriculture land use, while pastures and open forests on the side slopes and gully bed. Generally ravine areas have 15 to 32% arable lands and 60 to 85% non-arable lands [2]. The marginal lands adjacent to ravines are under extensive cultivation without much attention to conservation measures, which leads to further extension of gullies head during rainy season. These ravine lands are under biotic stress due to uncontrolled grazing by stray and wild animals and poor vegetation cover leading to huge soil loss during rainy season. These are the worst form of land degradation formed, when the vegetal cover is not strong enough to hold and bind the soil particles together from being carried away by the runoff water. The restoration of these ecosystems through bio-engineering measures is extremely important to conserve natural resources, maintain landscape sustainability, enhance carbon sequestration, mitigate climate change, improve socio-economic conditions and ensure food and livelihood security throughout such regions [3].

Ravine is subjected to extreme hydrological events (flood and drought), high summer temperatures, high wind velocity and extreme soil erosion that make them more vulnerable to land degradation. The several anthropological activities such as improper land use, illicit mining, faulty methods of road construction, uncontrolled grazing, deforestation, increased population, overexploitation of natural resources, unplanned urbanization, unregulated industrialization, increased rural poverty, low natural resource management skill and absence of appropriate resource conservation measures etc. accelerates the erosion process, that leads to land degradation and ravine landscapes formation [4]. To reclaim and rehabilitate such unproductive ravines, soil and water conservation technologies, e.g., peripheral and contour bund, vegetative barriers, grassed waterways, terrace, trench, composite check dams, gully plugs, gully easing, afforestation, agroforestry, contour cultivation, strip cropping, intercropping, mulching, tillage, cropping system, crop scheduling, crop geometry, organic manures, conservation agriculture, and soil management techniques have been successfully tested and recommended. These soil and water conservation measures can provide a large number of tangible i.e. controlling soil erosion, improving soil properties, promoting plant growth and yield, enhancing biomass/carbon stock, and provision of various tree products and intangible benefits i.e. mitigating climate change, protection of downstream water bodies, regulating the environmental flow, biomass recycling and soil formation, capacity building, out migration mitigation, bio-diversity [5]. Therefore, we have discussed the various engineering/mechanical, afforestation, agroforestry, agronomic and soil management practices for applied for managing and stabilizing the ravine slope.

2. Extent of ravine in India

In India the ravines have mostly developed along the river systems having highly productive and deep alluvial soils. The National Commission on Agriculture estimated 3.67 M ha of ravine lands 1.12% of total geographical area of India [6]. Out of 3.67 M ha of ravine lands in India, 2.36 M ha (64%) exists in Uttar Pradesh, Madhya Pradesh, Bihar, Rajasthan, and Gujarat states of India [7]. The ravines on *Yamuna* and *Chambal* River are the largest and most severely degraded ecosystems in the country. Besides these the ravines also extends from the banks of the *Tapti*, *Narmada*, *Watrak*, *Sabarmati*, and *Mahi* River. However, some staggered patches of ravines are also found in *Chota Nagpur* plateau, *Mahanadi* river basin and upper *Sone* Valley, *Indo-Gangetic* plains, *Shiwaliks* and *Bhabar* tract, and Western Himalayas even up to the *Kashmir* Valley [7]. In unmanaged conditions rate of ravine extension ranges from 0.6 to 1.0 m per year and is much greater on sandy soil than on the clay loam soils. The average annual loss of nutrients from these ravine lands due to soil erosion has been estimated at 5.37–8.4 Mg resulting into loss of production due to non-reclamation of ravines [8].

3. Classification of ravine system

The ravine system has gullies having independent catchment along with a regular main stream. In each drainage system, it is observed that gullies occur in

Symbol	Description	Specifications
G1	Very small gullies	Depth up to 3 m. Bed width not greater than 18 m. Side slopes varies.
G2	Small gullies	Depth up to 3 m. Bed width greater than 18 meters. Side slopes varies.
G3	Medium gullies	Depth between 3 m to 9 m. Bed width not less than 18 m. Sides uniformly sloping between 8 to 15%.
G4	Deep and Narrow gullies	Depth 3 m to 9 m. Bed width less than 18 m. Side slopes varies. Depth greater than 9 m. Bed width varies. Side slopes varies. Mostly steep or even vertical with intricate and active branch of gullies.

Table 1.
Classification of gullies in ravine land.

Land capability class	Slope (%)	Distance from gully rim and land form	Recommended soil and water conservation measures
a. Land suitable for cultivation			
I (light green yellow)	0–1	Very far from gully rim	No measures necessary, except ordinary good farming practices.
II (yellow)	1–3	Beyond 60 m	(i) Contour bund in the catchment, (ii) Good agronomic measures such as contour cropping, strip cropping and cover cropping.
III (red)	0–8	Between 6 and 60 m. Table lands and wide humps between wide gullies	(i) Contour and peripheral bund, (ii) Provision of safe outlets at gully heads (pipe outlet or chute) for the discharge of excess runoff, (iii) Intense agronomic measures such as contour cropping, contour strip cropping and cover cropping.
IV (blue)	8–15	Small and medium gully sides and beds (G1, G2, and G3). Slope must be uniform and gully depth should not be more than 9 m.	(i) Peripheral control with diversion bunds and safe outlets, (ii) Clearing and minor leveling on gully sides and beds, (iii) Putting up composite check dams of earth and brick masonry across the gully bed at 4′ vertical interval or 400 m horizontal interval, (iv) Terracing on gully sides up to 15% at 3′-4′ vertical interval, (v) Good crop rotations and heavy application of manure and fertilizers for restoration of soil fertility.
b. Land not suited for cultivation			
V (dark green)	15–25	Wide humps or gully sides (minimum 120 m long and 30 m wide with no gully intrusion.	Controlled grazing after pasture development
VI (orange)	Varies	Marginal land between the gully rim and the cultivated land (Peripheral bund)	(i) Grazing strictly prohibited, (ii) Cut the grass and stall feed the cattle's, (iii) Staggered contour trenching and afforestation.

Land capability class	Slope (%)	Distance from gully rim and land form	Recommended soil and water conservation measures
VII (brown)	Varies	Gully sides and beds (G4)	(i) Staggered contour trenching for moisture conservation and afforestation, (ii) Gully plugging and easing of the rapids along the gully beds, (iii) Development of pasture and afforestation.
VIII (purple)	Varies	Gully sides and beds (G4) with rock on the surface.	Complete closure to grazing and felling of trees.

Table 2.
Land capability classification for soil and water conservation measures in ravine land.

a certain regular order with well-defined side slopes, bed width and depth. In the upper reaches of the drainage system the gullies are wide and shallow with varying side slopes. The middle part of the drainage system usually has relatively deeper, wider and has uniform side slopes normally up to about 25%. The lower portion of the drainage system usually happens very deep, has steep side slopes and associated with intricate branched gullies. The ravine management depends on extent of gully bank deformation, slope, soil quality and vegetation cover. The ravines are classified in six classes based on the extent of gully bank deformation and erosion vulnerability under varying slopes. The shallow gullies up to 3.0 m depths are classified under land capability class I and II and recommended for cultivation of seasonal crops with moderate land leveling. The land capability class III has deeper and narrow gullies and more limitations for cultivation of seasonal crops. The land capability class IV has severe limitations of soil texture, gully size, steeper side slopes, and deeper and narrower gully beds for cultivation of seasonal crops and recommended for perennial horticulture plantations. The Class V and VI lands are not recommended for high value plantations or cultivation due to limitations of seasonal backflows from an adjoining river system, waterlogging, development of soil salinity due to irrigation, or due to extreme slope of gully bank. Hence, Class V and VI lands are recommended for perennial vegetation avoiding uncontrolled grazing activities. The ravine management is location-specific and should be undertaken based on the above classification and limitations.

Several workers have classified the gullies based on their cross section, forms, gully head characteristics, length, width and depth of catchment. Classification of the gullies, evolved at the ICAR-Indian Institute of Soil and Water Conservation, Research Centre, Vasad, Gujarat, India after critically observing ravine landscape developed along most of the river banks in Western India. The different types of gullies classified based on depth from adjacent marginal lands, width of the gully and side slopes given below in **Table 1** and the land capability classification and recommended conservation measures for ravine land is given in **Table 2** [9].

4. Measures for management of ravine land

The agricultural activity with land leveling is preferred for slope stabilization and its management in shallow ravines due to presence of deep and alluvial soil. The increased cost of land leveling operation is a major constraint and key factor in making decision for management of medium and deep ravines. The appropriate ways for management of ravines is reducing runoff intensity and safe disposal of runoff in ravines. The runoff generated from the adjacent marginal land is managed by constructing peripheral bund, checking of runoff/sediment in the interconnected gullies by earthen gully plugs/bori-bund/composite check dams at proper horizontal interval. The terracing is recommended in shallow and medium ravines slopes along with agroforestry measures for sustainable land use. The deep and very deep ravines are recommended to be kept under permanent vegetative cover along with site specific soil and water conversation measures. The medium and deep gullies having steeper slopes and economically not suitable for cultivation can be put under silvi-pasture permanent vegetation cover with protection from grazing for proper management of these lands.

4.1 Engineering measures for management of shallow ravine land

About one third of the ravine lands in the country are shallow ravines (<1 m deep) and can be easily reclaimed with simple earth moving machinery for cultivation of crops. The management of shallow ravines instigates with designing and instituting a peripheral bund along the gully head to check the runoff generated from the adjoining marginal lands. The land leveling operations across the slope and smoothening on the upstream side of the peripheral bund increases the infiltration opportunity time for the runoff generated from the crop field. The in-situ soil moisture conservation increases the water and nutrient use efficiency of the cropping systems.

4.1.1 Vegetative barrier

Vegetative barriers are used either for supplementing or substituting earthen bunds. The *Dichanthium annulatum, Cenchrus ciliaris, Vetiveria zizanioides, Eulaliopsis binata, Saccharum munja* and *Aloe barbadensis* vegetative barriers were evaluated for their effectiveness in reducing runoff, soil and nutrients losses from 2% slope of marginal shallow ravines. Vegetative barriers were grown across the slope at 45 m horizontal interval, in paired rows of 10 cm slip to slip spacing. The cultivation of pigeon pea (BDN-2) was done at 120 × 30 cm spacing in all plots. These vegetative barriers reduced the annual runoff by 19.7 to 50.1% and soil loss by 51.1 to 80.3% over the control plot [10].

The Napier, Guinea and Para fodder grasses strips of 1 m and 2 m were grown as vegetative barrier in 2% slope in shallow ravines. The lowest sediment yield was observed in 2 m width of Napier grass strip (1.43 t/ha) followed by 2 m width of Guinea and Para grass strips as compared to control (4.07 t/ha), respectively. The 2 m width of Napier grass strips has lowest nutrient loss of (N-5.64 kg/ha, P-3.1 kg/ha, K-5.4 kg/ha) followed by Guinea and Para were found almost equally efficient in reducing nutrient losses from crop field which is 28 to 30% of nutrient losses from control (N-20.01 kg/ha, P-8.3 kg/ha, K-17.00 kg/ha). The equivalent yield of 2 m width of Napier grass strip and cotton crop (1293.5 kg/ha) has highest yield followed by others. The economic analysis shows that Napier grass strips of 2 m width was found best in reducing runoff, soil loss, nutrient losses and net return from crop field [10].

4.1.2 Grassed waterway

Grassed waterways are important for preventing the scouring of channel bed in shallow ravines. The *Para* grass strips in waterways were optimized for grass cover to check runoff velocity and reduce sediment concentration in downstream water bodies. The different *Para* grass cover (0–100%) in waterways was studied at 2% slope. The different grass covers are 100%, 75%, 50%, 25% and no grass cover. These *Para* grass strips were able to produce green grass yield of 14.5 kg/m^2. The *para* grass strips in waterways is able to reduce the outflow up to 22% with 100%

grass cover. The grass filter strips is able to reduce the sediment concentration in runoff water by 5 times (from 3.6 to 0.72 g/lit). The *Para* grass filter in waterways is able to reduce the flow velocity of runoff water by converting the super critical flows into sub critical flows [10].

4.1.3 Contour and peripheral bund

The management of shallow ravine by contour and peripheral bund has shown improvement in value of land and crop yield due to soil moisture retention within the field, prevention of soil and nutrient losses from the crop field. The different cross sections of contour and peripheral bunds were tested in ravine lands of Western India [11]. The design cross sections of the bunds were fixed on the basis of the area of the catchment, slope of the land and its location. The bunds were designed on the basis of the vertical interval or horizontal spacing [11]. Usually one bund could only be located in each field approximately on contour so that it can also serve as peripheral control of gully heads. These bunds were sodded with *Dichanthium annulatum* and *Cenchrus ciliaris* grasses. In ravine lands an average cross section of 0.9 to 1.3 m^2 spaced at 0.90 to 1.20 m vertical interval was found best [11]. Grass ramps and pipe outlets were provided in the bunds for the safe disposal of excess runoff water. It is further stated that the effectiveness of the bunds would be for a longer period, as with passage of time the rate of reduction of bund height is expected to be slow and can be maintained indefinitely with good stabilized grasses [12]. It is also found that the area lost under bunds could fetch revenue from grasses, which is sufficient as comparable to production in rainfed area [12].

4.2 Engineering measures for management of medium and deep ravine land

About two third of the ravine lands in the country are deep ravines 3–9 m or more with varying width and slope and cannot be easily reclaimed with simple earth moving machinery for cultivation of crops. The management of deep ravines instigates with designing and instituting a series of composite check dams in the gully bed. The construction of gully plug at regular interval with provision for safe disposal of runoff assists in stabilization of gully bed, which can be alternatively utilized for raising crops tolerant to water logging. The easing of gully is required in deep gully to prevent caving action against steep slope due to runoff and protect the adjoining marginal land to collapse inside the gully due to unstable slope. The medium and deep gully can be reclaimed by terracing and or trenching for conservation of runoff, soil loss, associated nutrients and stabilize the steep slopes with time.

4.2.1 Composite check dam

A large number of earth cum brick masonry check dams were constructed in Mahi ravines of Western India for reclamation of medium and deep gullies in ravine [10]. These check dams were found to be very effective to check erosion, detaining the sediment and runoff water behind the structure which ultimately resulted in ground water recharge. The deposition of sediment against the check dams was measured by fixing a series of angle iron poles on concrete level of the structure. The average sedimentation from these ravine sub catchments having agricultural crop in table lands as well as in gully beds was 24.51 cu. m./ha/year during year 1961 to 1963 [10]. The average sediment deposition from watershed having agricultural crop in tablelands and natural regeneration in gully beds was 4.20 cu. m. /ha/year during year 1964 to 1977 [10]. After the siltation of these composite check dams the level terraces formed in the gully beds were stabilized and was put under cultivation but it was found to be poor in production. Therefore these reclaimed cultivated deep gully beds were subsequently planted with forest trees [10].

4.2.2 Gully plugs

Gully plugs protect the gully beds by reducing the runoff velocity, distributing the water spread, increasing infiltration opportunity time and improving the soil moisture regime for improving the vegetation cover. Gully plugs made of various materials i.e. brush wood, live hedges, earth, sandbags and brick masonry, loose boulder were evaluated in Western India [10]. The size and materials for the gully plug depends on the width, length and bed slope of the gully and anticipated runoff. In narrow gullies whose width did not exceed 3 m, live hedges consisting of Euphorbia species were planted across the gully beds in three rows spaced 90 cm apart and the stems at 90 cm in each row alternatively staggered [11]. It was found that all types of gully plugs were effective either in retaining or retarding the runoff. The earthen gully plugs were found to be the cheapest. Brick masonry gully plugs are constructed at the confluence of all gully branches of a compound gully. The gullies where no runoff is expected from the top, earthen gully plugs of 1.1 m^2 cross section with a grassed ramp of 22.5 cm below the top level and spaced at 45–60 m horizontal interval were found suitable. However, for gullies in which excess runoff from the top was expected an earthen gully plugs of 2.2 m^2 cross section with a pipe outlet was to be provided. The earthen gully plugs are required to be constructed for a life expectancy of 10 years. During these periods it is estimated that the vegetative growth of forest species will be sufficient to take care of soil erosion as well as their root system will be sufficiently developed to extract the moisture from deep soil layers [11].

4.2.3 Gully easing

Deep and vertical gully heads in association with the phenomena of under cutting or caving extending at an alarming rate endangering buildings, roads, bridges, abutments, railway tracks and costly cultivated lands. The measure which can be economically and immediately adopted to stop further progress of the gully head or cave the bottom of gully and then ease the gully for the remaining one third top portion of the vertical face. The newly formed slope of about 3:1 is stabilized by sodding with *Dichanthium annulatum* or *Cenchrus ciliaris* grasses [11]. The eased gully heads were found out to be in existence and working satisfactory even after 22 years of management. The comparison of eased gully with nearby gully under similar condition reveals that there is good vegetation on the eased gully head in comparisons to the unease gullies which are still exposed and covered with sparse annual vegetation [13].

4.2.4 Bench terracing

The impact of bench terracing on runoff, soil loss, and soil properties along with Sapota (*Achras zapota*) growth, fruit yield, biomass, and carbon stock in a degraded ravine land developed along the course of *Mahi* River in Western India were evaluated. The bench terracing in deep ravines with uniform slope of 15% resulted in significant decrease in runoff (34%) and soil erosion (25%), and enhanced tree growth, biomass and carbon stock. The cultivation of crops in between the tree plantation may induce significant soil loss (18% higher) due to tillage operations even though the runoff is not significantly affected. The findings suggested that bench terracing is the best soil and water conservation measure for restoring highly degraded ravines [3].

4.2.5 Trenching

The impact of trenching in deep ravines was evaluated ravines developed along the Mahi River in Western India. The staggered contour trenches were designed based on maximum daily rainfall. The trenching density on the ravine slopes was kept to retain 30, 50 and 80% of runoff generated from the ravine catchment. A substantial reduction in runoff was observed for treatments with higher trench densities (50% and 80%) as compared to 30%. The sediment yield in different trench densities also followed the similar trend. The soil moisture was more or less similar in different trench densities just followed by the monsoon. However, soil moisture in the 80% trench density was highest and also remains for a longer period in the lower reaches of the ravine slopes. The survival of Neem (*Azadirachta indica*) saplings planted at a spacing of 6 m × 6 m was also recorded and was also found highest in 80% trench density. In another study, the trenching of size 2 m × 0.5 m × 0.5 m at 14% uniform slope with Sapota plantation resulted in decrease in runoff by 16% and soil loss by 15% along with enhanced tree growth, biomass and carbon stock in the deep ravine slopes [11].

4.2.6 Bamboo based bio-engineering measures

The bamboo based bio-engineering measures is found as an effective means for natural resource conservation in ravines [14, 15]. The soil erosion is found comparatively less in bamboo plantations (178 kg ha^{-1}) compared with other forest plantation [16]. The significant stem flow and funneling ratio of bamboo plants provides better opportunity for rainfall absorption in degraded ravine lands [17]. Bamboo also acts as a vegetative barrier and filter for silt laden runoff flowing in gullies. The bamboo induces silt deposition and reduces the velocity of flowing water along bare river banks and deforested areas in ravines. *Dendrocalamus strictus* which occupies major area covered under bamboo plantation in India was found best for economic utilization of ravines. It is reported that annually about 4000 culms of bamboo per hectare can be harvested from ravine lands [18]. The benefit–cost ratio of 1.98 from bamboo plantation in ravines is profitable having an economic return of 19.3% over a period of 20 years [19]. The intangible benefits of carbon sequestration and prevention of soil erosion are supplementary. The ravine lands under the bamboo plantation increased the soil pH, organic carbon along with reduced runoff [20].

4.3 Forestry and agroforestry measures for management of ravines

The engineering measures are expensive and also needs technical acquaintance for proper designing and execution for management of ravines. The medium and deep gullies of ravine needs permanent vegetative cover having deep root system to bind the soil against strong forces of erosion in gullies during rainy season. The afforestation and agroforestry are relatively cheaper and affective in checking the soil erosion in medium and deep gullies. The performance of various afforestation and agroforestry system evaluated in ravine developed along the Mahi River in Western India is discussed below.

4.3.1 Afforestation

Afforestation is an important tool to mitigate land degradation due to soil erosion. Afforestation in ravine land provides ecosystem provisioning services i.e. fodder, timber, fuel, non-timber forest product, medicine and gum; regulating services i.e. nutrient cycling, controlling soil erosion, moderating climate, carbon sequestration; supporting service i.e. net primary production, soil formation and cultural services, recreation [3, 4]. The vegetation exploration in ravine shows the dominance of flora like *Azadirachta indica*, *Acacia nilotica* and *Prosopis cineraria*. In ravine gully head, there is predominance of *Prosopis cineraria*, *Acacia nilotica*, *Acacia senegal*, *Azadirachta indica*, *Holoptelia integrifolia*, *Feronia elephantum* and *Balanites roxbyrghii*. In ravine gully slope, there is equal distribution of *Acacia nilotica*, *Acacia senegal*, *Azadirachta indica* and *Prosopis cineraria*, followed by *Holoptelia integrifolia* and *Salmalia*. In bottom of ravine gully beds, there is predominance of *Holoptelia* species, *Acacia* species, *Azadirachta indica*, *Anogeissus latifolia* and *Balanites roxburghii*. Artificial regeneration results indicated the dominance of *Acacia nilotica* followed by the *Azadirachta indica*. The other dominant species were *Prosopis cineraria*, *Holoptelia integrifolia*, *Salmalia malabaricum* and *Anogiessus* species. *Acacia nilotica* attained diameter at breast height (DBH) between 19.6 cm and 28 cm in the Mahi ravines of Western India after 28 years. Semi-arid edaphically degraded soils in ravine lands are dominated by low value thorny trees and the plantation of well adapted and commercially important tree species is utmost important in such lands for obtaining higher ecological and economic benefits. In fact, the effective forest cover is less than 10%, causing shortage of fuel, fodder and timber wood in the area and further minimum 50% forest cover is required to prevent the land degradation. Therefore in such conditions maintenance of continuous canopy covers is extremely important to check soil erosion. *Eucalyptus tereticornis* and *Leuceana leucocephala* was found to be a fast growing tree species and is in much demand especially in the industries [10]. Plantation of bamboo (*Dendrocalamus strictus*) was found highly effective in enhancing vegetation cover, preventing soil erosion and stabilization of ravine slopes [21]. The uncontrolled grazing in ravines accelerates soil erosion which cannot be completely stopped. Therefore a system of restricted and rotational grazing may be helpful to some extent. In ravines, huge gap exists between the demand and supply for timber wood, and sustainable utilization of ravine land can reduce the demand–supply gap. Promotion of awareness and encouragement of tree planting on individual farm land can considerably improve the regional ecology and economy of the local people. These measures also help to improve soil carbon stock that may positively influences soil physical, chemical and biological property, improving the overall productivity of such ecosystem.

4.3.2 Agroforestry

Agroforestry is perennial and annuals plants combination that enhances landscape productivity and conserve natural resources. Agroforestry measures particularly for soil conservation have shown great potential in enhancing productivity of ravine lands [5, 22]. *Prosopis cineraria*, *Acacia leucophloea* and other regional multipurpose trees (MPT's) grown on field boundaries in Gujarat state of Western India. Sometime negative effect on crops may also occur due to competition for light, moisture and nutrients between the trees and crops. The combinations of *Dendrocalamus strictus*+*Cenchrus ciliarus* was also found successful in ravine slopes. *Dendrocalamus strictus*+*Cenchrus ciliarus* the reduced sediment yield 10–20 times less, compared to area without plantation [19]. *Moringa oleifera* and *Embilca officinalis* based agri-horticultural systems were also found effective in enhancing land productivity, conserving natural resources and improving soil health. Soil and water conservation measures (SWCM) also helps to improve agroforestry trees survival and growth due to improved soil moisture and reduced soil erosion. in an agroforestry experiment observed that Cowpea + Castor cultivation in between

Sapota trees may induce soil loss on ravine slope [3]. Supporting native or improved perennial grasses instead of crops may be a better option to reduce the soil erosion. Most researchers have concluded that incorporating trees in agriculture landscape conserve soil and water resources in ravine lands. In these conditions, extreme weather condition also affects plant growth and productivity and hence crop failure is a big problem. Tree-crop combinations system can be win-win situation in these climatically and edaphically venerable degraded agro-ecosystem. Tree and grass systems are also considered to be more resilient compared to crops during extreme weather conditions. Incorporating trees agriculture system also results in higher carbon stock and sequesters maximum carbon, compared to sole cropping systems. The CO_2 mitigation in agroforestry is approximately 11–41% higher compared to sole cropping systems. The farmers in ravine should shift from annual crops to perennial plantations to conserve soil resources, avoid risk of crop failure and mitigate climate change. Moreover, a properly managed agroforestry systems can moderate/improve climatic conditions, conserve soil and water resources, improve ecosystems services and absorb atmospheric GHG's in ravine lands. This suggested agroforestry based farming systems should be preferred globally in degraded ravine agro-ecosystems.

4.4 Agronomic measures for management of ravine

Agronomic measures are being considered as second line of defense in soil and water conservation however, these are of immense importance perhaps much more than other practices. These measures are recommended on mildly sloping lands (1–6%) for conservation of rainfall for sustainable production. They help in intercepting rain drops by decreasing the amount, intensity and the spatial distribution of the precipitation reaching the soil surface and protects the soil surface from the direct impact of raindrops which can cause a splash and sheet erosion. They also help in increasing infiltration rates and thereby reduce runoff and overland flow. These measures include contour cultivation, strip cropping, intercropping, mulching, tillage and other improved practices [23].

4.4.1 Contour cultivation

When tillage operations are carried out along the slope, the flow of generated runoff is accelerated, because each furrow serves as a rill which results in more runoff and soil erosion. Carrying out the farm operations such as plowing, seeding, planting and inter-culture operation along the contour or across the slope helps in formation of natural ridges and furrows which act as series of mini barriers and reservoirs to intercept rainwater and reduce runoff, soil and nutrient losses. In Shiwalik foothills, contour cultivation reduced runoff by 20.6%, soil loss by 43.5% and increased maize yield by 23% [24].

4.4.2 Strip cropping

It is the system of growing of erosion permitting crops (Maize, Sorghum, Pearl millet, Cotton etc.) and erosion resisting crops (Green gram, Black gram, Moth, Groundnut etc.) in alternate strips of suitable width across the slopes on contour in the same field. It helps in reducing the slope length, reducing velocity of runoff, arresting soil by provide a vegetative filter and increasing opportunity time for infiltration of rainwater into the soil profile. In Doon valley, strip cropping of Maize and Cowpea in 2:1 ratio has been reported to reduce runoff from 43 to 37% and soil loss from 21 to 11 Mg/ha [25].

4.4.3 Intercropping

The growing of low stature legumes like Groundnut, Cowpea, Green gram and Black gram in wider inter-row spaces of crops like Maize, Sorghum and Castor simultaneously on the same piece of land with a definite row-planting pattern is known as intercropping. It provides sufficient cover on the ground and thereby reduces soil erosion in shallow ravines. Intercropping offers an opportunity for profitable utilization of available space. Intercropping of Castor + Green gram (1:2) in South-Eastern Rajasthan reduced runoff by 22% and soil loss by 30% as compared to sole castor, apart from improving soil fertility status and improving crop yield. The system has been found ideal under aberrant conditions such as late onset of monsoon where it recorded only 15% reduction in productivity against 48–50% reduction in other systems [26].

4.4.4 Mulching

It is one of the most important agronomic practices which not only reduces evaporation losses, runoff, and soil erosion but also increases infiltration, improves of soil structure and regulates of soil temperature. In low rainfall areas, mulching helps in conserving moisture in the soil profile while in high rainfall areas, it reduces runoff and soil losses, in turn, are reflected in higher productivity. A study conducted at Chandigarh in Shiwalik region of Northern India indicated possibility of increasing Wheat yield by 58% by providing grass mulch during *Rabi*. In another study conducted at Dehradun, recycling of Sun hemp *in-situ* as mulch one month after sowing reduced runoff and soil loss and significantly increased yield of succeeding Wheat crop [24]. At Kanpur in Northern India, rice straw mulch increased yield of maize by 41.8% and mustard by 190%. Mulching can reduce erosion on slopes of 25–39%. In tea garden the mulching with Guatemala grass (*Tripsacum laxum*) at the rate of 37 Mg/ha reduced the soil loss from 9 Mg/ha to 7 Mg/ha [27].

4.4.5 Conservation tillage

Tillage modifies the soil physical characteristics i.e. bulk density, surface roughness, porosity and hardness of pan. The conventional tillage leaves no land unplowed and leaves no residues on the field. In the Indo-Gangetic plains of Yamuna river, where rain water cannot enter into deeper layers due to presence of impervious soil layer (*Kankar*) and is lost as runoff, deep tillage by disc plow once in three years up to 20 cm depth and one pass of cultivator every year helped in reducing runoff considerably and enhanced grain yield of pearl millet by 50%. Conservation tillage involves disturbing the soil to the minimum extent necessary and leaving crop residues on the soil surface. It ensures at least 30% coverage of the soil surface with crop residue which play a very important role in organic carbon build up and soil and moisture conservation under dry land. In Maize-Toria cropping system of the hill and mountain agro-ecosystem, Minimum Tillage + Crop residue incorporation is effective in reducing runoff by 12% and soil loss by 24% as compared to the conventional tillage besides minimizing nutrient losses [28].

4.4.6 Cropping system

Soil erosion depends on cropping system adopted on the land. The growing of crops which produce maximum cover, reduce runoff and soil loss always better for soil erosion point of view. Cowpea and Green gram are important cover crops for the rainy seasons. They provide early and dense (85%) ground cover which generally coincides with peak rate of runoff. The splash erosion loss in fallow land is 44.2 Mg/ha while it is 26.3 Mg/ha in Green gram and Black gram plots. In a study at Gujarat in Western India it was found that *Bidi* Tobacco, being a clean cultivated crop, allows higher runoff and soil loss and these losses can be reduced by introducing cover cum green manure crops during the early monsoon before Tobacco is transplanted in the field [10].

4.4.7 Crop scheduling

Sowing date and optimum seed rate are pillars for good stand and crop cover. The sowing date must be so adjusted that by the time intensive rainfall take place, there is enough of vegetation cover on the land. This can be done by advancing the date of sowing or even dry sowing in rainfed areas is always recommended. A study conducted at Dehradun in Northern India reported that sowing of Maize at the start of last week of June month produced higher and dense canopy at peak rainfall period, which in turn reduced the splash erosion considerably as compared to crop sown on first week of July. Similarly, the canopy coverage and crop yields were higher under early sowing in Maize + Cowpea intercropping system [28].

4.4.8 Crop geometry

Optimum plant population and crop geometry are equally important to have good crop cover on the land. Narrow intra-spacing (within rows) across the slope offers tolerance to runoff and ultimately reduces the soil erosion. Narrow spacing is always better for soil and water conservation, but it may cause severe competition for nutrient, water, light and space resulting in in lower yield of crops. A study carried out Chandigarh in Northern India reported that growing of Maize at 60 cm × 22.5 cm with minimum tillage reduced the runoff from 281 to 253 mm, soil loss from 5.7 to 4.4 Mg/ha as compared to conventional crop geometry [29].

4.5 Soil management measures for ravines

In order to meet the nutrient requirements of the crops in ravines, the soil fertility loss due to extreme erosion needs to be compensated accordingly as per specific site. The steep decline in soil fertility in the absence of proper measures to check erosion will cause reduction in the crop yields. This can be achieved in the following manner.

4.5.1 Judicious use of fertilizers

The soil fertility in ravine land varies from gully head to gully bed, therefore fertilizer recommendation for crops and cropping system should be site specific based on soil test. The recommended dose of fertilizers, method of application and proper scheduling is required to achieve sustainable yield from ravine lands. The nitrogen fertilizers should be applied in splits basal dose at the time of sowing and as top dressing during crop growth to the crops. The phosphate and potash fertilizers are given only as basal doses. Spraying of liquid Urea and micronutrients should be done on the standing crop, as and when deficiencies are noticed in the crop. The selection of appropriate fertilizers and their accurate combinations are necessary for maintaining the soil fertility in ravines.

4.5.2 Organic manures

Organic manures such as FYM (Farm Yard Manure), composts and green manures improve the physical condition, micro-flora of the soil and have a beneficial residual effect on succeeding crops. It is therefore necessary that organic manure should be applied to maintain soil health and sustain crop yields on a long term basis. It is generally felt that adequate amounts of these manures are many times not available. Therefore it is suggested that green manures crops may be incorporated in the prevailing cropping systems periodically.

4.5.3 Bio-fertilizers

Bio-fertilizers in the context of present day organic farming can play a key role in sustainable agriculture because of their potential to restore soil health. They are derived from the living organisms and include nitrogen fixing microorganisms i.e. *Rhizobium* spp., *Azospirillum* spp., *Bradyrhizobium* spp., *Azotobacter* spp., *Frankia* spp., Phospho-microorganisms, i.e., phosphate solubilizing bacteria, *Glonus* spp., phosphate solubilizing fungi compost inoculant i.e., *Cellolotic* or *Lignolytic* fungi etc., bio-fertilizers provide better crop yields and are ecofriendly. Bio-fertilizers can be applied through seedling treatments; tuber set treatment or oil treatment. For 10 kg of medium size seeds 483 g of culture is required for one hectare of land. In case of seedlings 1–2 kg of bio-fertilizers may be dissolved in 10 liters of water in a bucket and a suspension/slurry can be prepared for application in soil. The roots of the seedlings are dipped for 15 minutes in the slurry and then transplanted in the field. For treatment of tubers or sets slurry of 2–4 kg bio-fertilizers is prepared in 40–80 liters of water. Before sowing, tubers or sets are dipped in the slurry for 15 minutes. Soil treatment with the bio-fertilizers can also be done in the standing crop at critical stages [29].

4.5.4 Vermicomposting

Vermicomposting is technique of using different species of earthworms to convert the decomposable farm, urban and domestic waste into nutrient rich compost. It is eco-friendly, sustainable, cheaper and easy technique to obtain compost of high quality using biologically decomposable organic waste and crop residue generated from the farmers field.

4.5.5 Integrated nutrient management

The chemical fertilizers are nowadays becoming costly and may not be timely available to the farmers in remote areas. Though chemical fertilizers increase yields quickly, yet their effects on soil is visible for a short period of time and requires frequent application to maintain the soil fertility. These chemical fertilizers have a toxic and residual effect in soil, water and plants system. In order to overcome these constraints, a conjunctive use of organic and inorganic sources to maintain soil fertility has been suggested. This is generally referred to as Integrated Plant Nutrient Management (IPNM). Under this scheme, the some portions of crop nutrient requirement are met by organic manures and the remaining need is fulfilled by the chemical fertilizers. In one of the studies, 20 kg N ha^{-1} by *Azolla* (a fern) + 20 kg N by ammonium sulfate applied to rice registered a grain yield of 4435 kg ha^{-1} in comparison to 40 kg N ha^{-1} with ammonium sulfate giving a yield of 4132 kg ha^{-1} which explains the role and efficacy of nitrogen fixation in the soil and its use by rice crop. *Rhyzobium* inoculation of the seeds of legumes is being promoted for building up soil nitrogen through nitrogen fixation by increasing the efficiency of legume crop. At the farm level, domestic farm and agro-industrial wastes arising out of the food grain, crop residue and fruit processing can be diverted into the soil. This will add to soil fertility as well as biological waste management [29].

5. Conclusion

The India is one of the first countries to commit to the United Nation Convention to Combat Desertification Sustainable Development Goal to achieve the land degradation neutrality by 2030 during COP-14 held at New Delhi in September, 2019. India will restore an additional 5 M ha of degraded land by 2030, raising the land to be restored in India to 26 M ha. Therefore, ravine ecosystem can become an important milestone in achieving the food and nutritional security in coming decades through reclamation of ravine land with different site specific interventions as discussed above. There is a great opportunity for providing sustainable livelihood to the resource poor people of these areas though reclamation of the ravine with people's participation. An integrated approach with production system for bio-process industry will also generate employment opportunities for local people. In addition to this ravine reclamation is expected to mitigate climate change through carbon dioxide sequestration. Therefore, ravine reclamation should need to be on priority for the policy makers to achieve the future sustainability goals.

Acknowledgements

The authors are thankful to all the Head, Scientist and Technical staff currently working or previously employed at ICAR-Indian Institute of Soil and Water Conservation, Research Centre-Vasad, Anand, Gujarat, India for providing the necessary data and resources to complete this book chapter. We are also thankful to Director, ICAR-Indian Institute of Soil and Water Conservation, Dehradun, India for his kind support and encouragement throughout this period.

Author details

Gaurav Singh[1*], Raj Kumar[2], Dinesh Jinger[1] and Dinesh Dhakshanamoorthy[1]

1 ICAR-Indian Institute of Soil and Water Conservation, Research Centre-Vasad, Anand, Gujarat, India

2 ICAR-Central Soil Salinity Research Institute, Karnal, Haryana, India

*Address all correspondence to: gaurav.bhu09@gmail.com

References

[1] Soni, M.L., Subbulakshmi, V., Renjith, P.S., Dagar, J.C., Yadava, N.D. (2018). Reclamation of Ravine Lands for Higher Production. In Ravine Lands: Greening for Livelihood and Environmental Security. pp. 279-307. Springer, Singapore.

[2] Pande, V.C., Kurothe, R.S., Singh, H.B., Rao, B.K., Kumar, G., Bhatnagar, P.R. (2018). Socio-economic and Conservation Measures in Ravine-affected Areas of Gujarat: Policy Interventions. In Ravine Lands: Greening for Livelihood and Environmental Security. pp. 591-600. Springer, Singapore.

[3] Kumar, R., Bhardwaj, A.K., Rao, B.K., Vishavkarma, A.K., Bhatnagar, P.R., Patra, S., Kumar, G., Kakade, V., Dinesh, D., Pande, V.C., Singh, G., Dobhal, S., Sharma, N.K. (2020a) Development of degraded ravine lands of Western India via Sapota (Achras zapota) plantation with terracing vs. trenching on slope based conservation measures. Land Degradation & Development. DOI: https://doi.org/10.1002/ldr.3691.

[4] Kumar, R., Bhatnagar, P.R., Kakade, V., Dobhal, S. (2020b). Tree plantation and soil water conservation enhances climate resilience and carbon sequestration of agro ecosystem in semi-arid degraded ravine lands. Agricultural and Forest Meteorology, 282, 107857.https://doi. org/10.1016/j. agrformet.2019.107857.

[5] Chaturvedi, O. P., Kaushal, R., Tomar, J. M. S., Prandiyal, A. K., & Panwar, P. (2014). Agroforestry for wasteland rehabilitation: mined, ravine, and degraded watershed areas. In Agroforestry Systems in India: Livelihood Security & Ecosystem Services. pp. 233-271. Springer, New Delhi.

[6] GoI (1976) Report of the National Commission on Agriculture 1976. National Commission on Agriculture, Government of India, Ministry of Agriculture and Irrigation, New Delhi.

[7] Dhruva Narayana V.V. (1993) Soil and water conservation research in India. Indian Council of Agricultural Research, Pusa, New Delhi. p. 454.

[8] Sikka, A.K., Mishra, P.K., Singh, R.K., Rao, B.K., Islam, A. (2018). Technological Interventions for Managing Ravine Lands for Livelihood and Environmental Security. In Ravine Lands: Greening for Livelihood and Environmental Security. pp. 217-236. Springer, Singapore.

[9] Tejwani, K.G. and Dhruva Narayana, V.V. (1961). Soil conservation survey and land use capability planning in the ravine lands of Gujarat. Journal of the Indian Society of Soil Science, 9(4), 233-244.

[10] Kurothe, R.S., Singh, H.B., Tiwari, S.P., Pande, V.C., Bagdi, G.L., Sena, D.R., Vishvakarma A.K. & Kumar, G. (2013). Fifty years of research in soil and water conservation. CSWCRTI, Research Centre, Vasad, India, pp. 10-100.

[11] Tejwani, K.G., Gupta S.K. and Mathur, H.N. (1975). Soil and water conservation research in India. (1956-71) I.C.A.R. Publication, New Delhi, India, pp. 254-310.

[12] Singh, B. and Verma, B. (1971) A comparative study of the economics of various soil conservation-cum-grass land improvement practices for rejuvenating forage production in ravine lands, (Forage production). Indian Forester, 97 (6): 315-321.

[13] Kurothe, R.S., Samra, J.S., Samarth, R.M. (1999). Optimum design of contour bunds for Navamota watershed (Gujarat)–a case study. Indian J. Soil Conserv, 27(1), 17-21.

[14] Lawler, D.M. (1993). The measurement of river bank erosion and lateral channel change: a review. Earth surface processes and landforms, 18(9), 777-821.

[15] Wang, Y., and Liu, Y. (1995). Hydrological characteristics of a moso-bamboo (*Phyllostachys pubescens*) forest in south china. Hydrological Processes, 9(7), 797-808.

[16] Ben-Zhi, Z., Mao-Yi, F., Jin-Zhong, X., Xiao-Sheng, Y., Zheng-Cai, L. (2005). Ecological functions of bamboo forest: research and application. Journal of Forestry Research, 16(2), 143-147.

[17] Rao, B.K., Kurothe, R.S., Pande, V.C., Kumar, G. (2012). Throughfall and stemflow measurement in bamboo (Dendrocalmus strictus) plantation. Indian Journal of Soil Conservation, 40(1), 60-64.

[18] Rao, B.K., Kurothe, R.S., Singh, A.K., Parandiyal, A.K., Pande, V.C., Kumar, G. (2012). Bamboo plantation based technological interventions for reclamation and productive utilization of ravine lands. CSWCRTI. Technical Bulletin No, 30.

[19] Pande V.C., Kurothe R.S., Rao B.K., Kumar G., Parandiyal A.K., Singh A.K., Kumar A. (2012). Economic analysis of bamboo plantation in three major ravine systems of India. Agricultural Economics Research Review 25(1): 49-59.

[20] Singh, A.K., Kala, S., Dubey, S.K., Rao, B.K., Gaur, M.L., Mohapatra, K.P., Prasad, B. (2014). Evaluation of bamboo based conservation measures for rehabilitation of degraded Yamuna ravines. Indian Journal of Soil Conservation, 42(1), 80-84.

[21] Rao, B.K., Pande, V.C., Kurothe, R.S., Singh, A.K., Parandiyal, A.K. (2018). Bamboo-based Bioengineering Interventions for Rehabilitation of Ravines. In Ravine Lands: Greening for Livelihood and Environmental Security (pp. 397-412). Springer, Singapore.

[22] Singh, B., and Singh, G. (2015). Biomass production and carbon stock in a silvi-horti based agroforestry system in arid region of Rajasthan. Indian Forester, 141(12), 1237-1243.

[23] Jinger, D. and Kakade, V.D. (2019). Land degradation and its management through soil and water conservation measure on arable land. Kerala Karshakan 7(5): 12-20.

[24] Sharda, V.N. and Juyal, G.P. (2016). Handbook of Agriculture. Conservation techniques for sustaining natural resources. pp. 323-370, DKMA, ICAR, New Delhi.

[25] Ghosh, B.N., Sharma, N.K. and Dadhwal, K.S. (2011). Integrated nutrient management and intercropping/cropping system impact on yield, water productivity and net return in valley soils of north-west Himalayas. Ind. J. Soil Conserv. 39: 236-242.

[26] Rao, J.V. and Khan, I.A. (2003). Research Gaps in Intercropping Systems under Rainfed Conditions in India, an On Farm Survey; CRIDA: Hyderabad, India.

[27] Reddy, T.L. and Reddy, G.H. (2010). Principles of Agronomy (Fourth revised edition). Soil Conservation. pp. 310-336, Kalyani Publication, New Delhi.

[28] Sharma, N.K., Ghosh, B.N., Khola, O.P.S. and Dubey, R.K. (2013). Residue and tillage management for soil moisture conservation in post maize harvesting period under rainfed conditions of north-west Himalayas. Ind. J. Soil Conserv. 42:120-125.

[29] Dhruva Narayan, V.V. (2017). Soil and water conservation research in India (Third edition). Agronomic measures in soil and water conservation. pp. 153-174, Indian Council of Agricultural Research, New Delhi.